MONOGRAPHS ON
STATISTICS AND APPLIED PROBABILITY

General Editors

D.R. Cox, D.V. Hinkley, N. Reid, D.B. Rubin and B.W. Silverman

1 Stochastic Population Models in Ecology and Epidemiology *M.S. Bartlett* (1960)

2 Queues *D.R. Cox and W.L. Smith* (1961)

3 Monte Carlo Methods *J.M. Hammersley and D.C. Handscomb* (1964)

4 The Statistical Analysis of Series of Events *D.R. Cox and P.A.W. Lewis* (1966)

5 Population Genetics *W.J. Ewens* (1969)

6 Probability, Statistics and Time *M.S. Bartlett* (1975)

7 Statistical Inference *S.D. Silvey* (1975)

8 The Analysis of Contingency Tables *B.S. Everitt* (1977)

9 Multivariate Analysis in Behavioural Research *A.E. Maxwell* (1977)

10 Stochastic Abundance Models *S. Engen* (1978)

11 Some Basic Theory for Statistical Inference *E.J.G. Pitman* (1979)

12 Point Processes *D.R. Cox and V. Isham* (1980)

13 Identification of Outliers *D.M. Hawkins* (1980)

14 Optimal Design *S.D. Silvey* (1980)

15 Finite Mixture Distributions *B.S. Everitt and D.J. Hand* (1981)

16 Classification *A.D. Gordon* (1981)

17 Distribution-free Statistical Methods *J.S. Mariz* (1981)

18 Residuals and Influence in Regression *R.D. Cook and S. Weisberg* (1982)

19 Applications of Queueing Theory *G.F. Newell* (1982)

20 Risk Theory, 3rd edition *R.E. Beard, T. Pentikainen and E. Pesonen* (1984)

21 Analysis of Survival Data *D.R. Cox and D. Oakes* (1984)

22 An Introduction to Latent Variable Models *B.S. Everitt* (1984)

23 Bandit Problems *D.A. Berry and B. Fristedt* (1985)

24 Stochastic Modelling and Control *M.H.A. Davis and R. Vinter* (1985)

25 The Statistical Analysis of Compositional Data *J. Aitchison* (1986)

26 Density Estimation for Statistical and Data Analysis *B.W. Silverman* (1986)

27 Regression Analysis with Applications *B.G. Wetherill* (1986)

28 Sequential Methods in Statistics, 3rd edition *G.B. Wetherill* (1986)

29 Tensor methods in Statistics *P. McCullagh* (1987)

30 Transformation and Weighting in Regression *R.J. Carroll and D. Ruppert* (1988)

31 Asymptotic Techniques for Use in Statistics *O.E. Barndoff-Nielson and D.R. Cox* (1989)

32 Analysis of Binary Data, 2nd edition *D.R. Cox and E.J. Snell* (1989)

33 Analysis of Infectious Disease Data *N.G. Becker* (1989)

34 Design and Analysis of Cross-Over Trials *B. Jones and M.G. Kenward* (1989)

35 Empirical Bayes Method, 2nd edition *J.S. Maritz and T. Lwin* (1989)

36 Symmetric Multivariate and Related Distributions *K.-T. Fang, S. Kotz and K. Ng* (1989)

37 Generalized Linear Models, 2nd edition *P. McCullagh and J.A. Nelder* (1989)

38 Cyclic Designs *J.A. John* (1987)

39 Analog Estimation Methods in Econometrics *C.F. Manski* (1988)

40 Subset Selection in Regression *A.J. Miller* (1990)

41 Analysis of Repeated Measures *M. Crowder and D.J. Hand* (1990)

42 Statistical Reasoning with Imprecise Probabilities *P. Walley* (1990)

43 Generalized Additive Models *T.J. Hastie and R.J. Tibshirani* (1990)

44 Inspection Errors for Attributes in Quality Control *N.L. Johnson, S. Kotz and X. Wu* (1991)

45 The Analysis of Contingency Tables, 2nd edition *B.S. Everitt* (1992)

46 The Analysis of Quantal Response Data *B.J.T. Morgan* (1992)

47 Longitudinal Data with Serial Correlation: A State-Space Approach *R.H. Jones* (1993)

48 Differential Geometry and Statistics *M.K. Murray and J.W. Rice* (1993)

49 Markov Models and Optimization *M.H.A. Davies* (1993)

50 Chaos and Networks: Statistical and Probabilistic Aspects *Edited by O. Barndorff-Nielsen et al.* (1993)

51 Number Theoretic Methods in Statistics *K.-T. Fang and W. Yuan* (1993)

52 Inference and Asymptotics *O. Barndorff-Nielsen and D.R. Cox* (1993)

53 Practical Risk Theory for Actuaries *C.D. Daykin, T. Pentikainen and M. Pesonen* (1993)

54 Statistical Concepts and Applications in Medicine *J. Aitchison and I.J. Lauder* (1994)

55 Predictive Inference *S. Geisser* (1993)

56 Model-Free Curve Estimation *M. Tarter and M. Lock* (1993)

57 An Introduction to the Bootstrap *B. Efron and R. Tibshirani* (1993)

(Full details concerning this series are available from the Publishers.)

Model-Free Curve Estimation

MICHAEL E. TARTER
MICHAEL D. LOCK

CHAPMAN & HALL
New York • London

This edition published by
Chapman & Hall
One Penn Plaza
New York, NY 10119

Published in Great Britain by
Chapman & Hall
2-6 Boundary Row
London SE1 8HN

Printed in the United States of America

Library of Congress Cataloging in Publication Data

Tarter, Michael E.

Model-free curve estimation / Michael E. Tarter, Michael D. Lock.
p. cm.

Includes index.
ISBN 0-412-04251-7-5
1. Estimation theory. 2. Curve fitting. I. Lock, Michael D.
II. Title.
QA276.8.T36 1994
519.5' .44—dc20 93-38411
CIP

British Library Cataloguing in Publication Data available

Please send your order for this or any **Chapman & Hall book to Chapman & Hall, 29 West 35th Street, New York, NY 10001, Attn: Customer Service Department.** You may also call our Order Department at 1-212-244-3336 or fax your purchase order to 1-800-248-4724.

For a complete listing of Chapman & Hall's titles, send your requests to **Chapman & Hall, Dept. BC, One Penn Plaza, New York, NY 10119.**

Contents

Preface

Before 1956 statistical techniques such as the histogram and life-table were considered conceptually distinct from procedures like Gram-Charlier-Edgeworth Expansions and Pearson Family curves. Curves as parts of graphs and curves as functions were treated very differently. Books which dealt extensively with histogram and stick-pin map construction, such as Brinton's (1917) classic, *Graphical Methods for Presenting Facts,* usually contained no formulas and never mentioned such staples of the statistician's tool-kit as maximum likelihood estimation or even the normal curve. Brinton's book devotes 14 pages to a chapter entitled "Cumulative Curves" and 43 pages to a chapter entitled "Frequency Curves. Correlation." In neither of these chapters is there a single formula or equation.

Today the field commonly called curve estimation is making up for lost time. Instead of containing elementary yet picturesque graphs, such as Figure 1 below from Hopkins and Bond (1905), books on this subject begin with what can seem to data analysts as a formidably high level of mathematical and statistical abstraction. It is therefore small wonder that the consumer of modern statistics tends to steer clear of the field covered by our book. After all, the histogram and the life-table have certainly stood the test of time. Yet even for the seemingly humble histogram, the work of Scott (1979, 1985, 1988, 1992), Scott and Factor (1981), Scott and Sheather (1985), and Scott, Tapia, and Thompson (1977), and many others have transmuted the lowly histogram into one of the most useful of the many new approaches described in this book.

Histograms, life-tables, Gram-Charlier-Edgeworth Expansions and Pearson General Family curves all can be treated as ancestors of some these new approaches. Largely because of this varied parentage, these new procedures are called by many names. For example, the term "nonparametric pdf estimation" is commonly used today when the particular curve dealt with is the same curve estimated by the ordinary histogram. The term "curve estimation" is used to stress the capacity to deal

Picturesque graph which might possibly contain some useful information, although this seems unlikely.

with curves more complex than that targeted by the histogram, while the term "model-free curve estimation" is used to emphasize the positive side of the negative designation, nonparametric.

Because the roots of the field described in our book are so deep and diverse, no single term has been universally accepted by all researchers whose work parallels the topics dealt with in this book. Our own personal preference, however, is for the two words *generalized representation*, because this term underscores the freedom and flexibility that we feel is the hallmark of exploratory curve estimation.

Unfortunately even this phraseology may be slightly misleading, because the word *generalized* can suggest a high degree of abstraction. For this reason, we will begin in the first chapter to explain precisely why it is that the concept of generalization, rather than any other single concept, is now considered of fundamental interest, not only to researchers who study the foundations of mathematical statistics, but to the practicing statistician and data analyst whose work rests on these foundations.

Chapters 1 through 8 depart radically from the style of most statistics books. We will introduce the topics which underlie exploratory curve estimation by drawing extensively on the early historical literature of both statistics and applied mathematics. Conversely, we will only draw upon the recent extensive literature of curve estimation when the topics dealt with in this book closely match some topic directly covered in a particular chapter. There are now several excellent texts and monographs that describe kernel, average shifted histogram, L_1 metrics, and penalized likelihood approaches. Whenever possible, we will indicate which one or more of these books deals with material that is relevant to the particular exploratory curve estimation topic discussed in this book.

From a historical overview the same fundamental change that is currently being brought about by new practical ways of making use of generalized statistical representation had occurred, not only once, but twice during the development of applied and theoretical mathematics; first in 1637, the publication date of René Descartes' *La Geometrie*, and then again in 1807, the date of Joseph Fourier's first submission of his basic paper on heat conduction to the Academy of Sciences of Paris. While 1637 and 1807 are now universally accepted as being watershed years for quantitative science, it remains to be seen to what degree the years 1951, in which F. Fix and J. L. Hodges' paper of nonparametric discriminant rules, or 1956, in which M. Rosenblatt's paper on generalized histogram were published, will be viewed as being similar to the years 1637 and 1807.

In the 1950s and 1960s enough of the key elements of generalized statistical methodology were put together to make the new approaches clearly competitive with model-based alternatives. Previously, histograms were distinguished by being called descriptive rather than inferential procedures. One entire chapter of this book is devoted to the applications of the histogram's progeny to Bayesian, rank-based, and other forms of inference. Attempts at enhanced representational generality, like Gram-Charlier and Edgeworth Expansions, as well as the Pearson Family, were treated only as extensions of models. As discussed in Chapter 2, today these early approaches can be seen to have as much in common with the histogram as with any particular model. This change in orientation underscores a departure from what had at one time been thought of as two widely separated statistical model-based and nonparametric continents. Instead, much of the subject matter of this book describes an isthmus whose width and scope is only beginning to be appreciated.

Yet enough is known today so that whenever possible we can distinguish between such classifications of curve estimation methodology as kernel and series representation, and mean integrated square error as compared to penalized likelihood criteria. In many circumstances the close resemblance between topics will also be underscored.

We will also emphasize practical implications of methodological differences and similarities. In particular, Chapters 8 and 9 deal with the practical ramifications of the term *curve* estimation as opposed to the more restrictive term *pdf estimation*. Hence, in these two chapters the word *generality* refers to different classes of curves, for example, pdf's, cdf's, survival, regression, Q-Q, and even location parameter likelihood functions.

We hope that we have aroused the reader's curiosity. After spending many years investigating this area, we have come to feel much the same as Ali Baba and his companions had been said to have felt when they found themselves in a cave where they first walked "about in ecstasy marvelling at the treasures" (Burton, 1885). At first there is a tendency to want to revel by oneself. However, before long there comes the realization that so much encompassed in this area that it would be a pity not to share one's rapture.

We would like especially to thank Christina C. Mellin whose constant guidance and wise comments formed the foundation upon which our efforts were based, William R. Freeman who made many computational and conceptual contributions to our work, as well as Florence R. Morrison, Martha Schumway and Dolores Lysmer who desplit our infinitives and polished our prose in many additional ways. B. Shube, E. Johnston, and later J. Kimmel gave us the initial impetus and later guided the final polishing phases of this book. Thanks also are due to Marthana Hjortland, Elizabeth Margosches, Americo Rivera, Abraham Silvers, Frank J. Massey Jr., Richard A. Kronmal, Nancy Mann, Jansin T. Wong-Lee, Ai-Chu Wu, Leonard Buchner, Steve Ullery, Mary Lock, David Hoel and George S. Malindzak who provided us with help, including our NIEHS grant RO1 ES05379 and the many other grants and contracts that have supported the research described below.

CHAPTER 1

Introduction to curve estimation

1.1 Example: Model identification

One of the best ways to introduce the field of curve estimation is to show what it can do. Following is a re-analysis of two data sets originally published by W.F.R. Weldon in an 1893 paper that may have been the first ever to advocate statistical analysis as a primary method for studying biological problems (Pearson, 1906). In Weldon's time, few biologists had any mathematical training and few considered quantitative techniques especially applicable to their subject. For example, species were classified on the basis of the description of one animal that was thought to be normal or typical, along with a few notes on other individuals exhibiting unusual characteristics. Weldon, on the other hand, was a firm believer in the new discipline of biometry. In his own words, which began the first issue of the journal *Biometrika*, Weldon states that inquiries into evolutionary biology must be based "not [on] an individual but a race, or a statistically representative sample of a race" (Weldon, 1901).

With this idea in mind, Weldon set to work collecting and measuring 2000 female shore crabs, half from Plymouth Sound in England and half from the Bay of Naples. On each crab Weldon made 11 different measurements, mostly concerning the dimensions of the carapace. Table 1.1 summarizes the data from one of the measurements, the right antero-lateral margin of the carapace, from the sample of crabs from Naples Bay. The data are expressed as percentages of the total carapace length and have been rounded off to the nearest .004.

To aid in his data analysis, Weldon used a graphical tool much like the histogram shown in Figure 1.1. Although routinely employed by statisticians and other scientists, histograms are rarely appreciated for what they really are—curve estimators. Specifically, histograms are estimators of the probability density function. (Although not strictly identical in the mathematical sense, the terms curve and function will often be used inter-

Table 1.1. *Weldon's (1893) measurements of the right antero-lateral margin of the carapace of 999 female crabs taken from the Bay of Naples.*

Length in thousandths of total carapace length	Number of individuals	Length in thousandths of total carapace length	Number of individuals
716–719	1	756–759	126
720–723	3	760–763	82
724–727	5	764–767	72
728–731	11	768–771	41
732–735	40	772–775	28
736–739	55	776–779	8
740–743	98	780–783	7
744–747	121	784–787	0
748–751	152	788–791	0
752–755	147	792–795	2

changeably throughout this book, except where there is a possibility of confusion.) While commonly too coarse to provide detailed information, histograms can reveal gross characteristics of the density such as extreme skewness or multimodality. From his analysis, Weldon was able to conclude that the true distribution of the measurements was symmetric. In

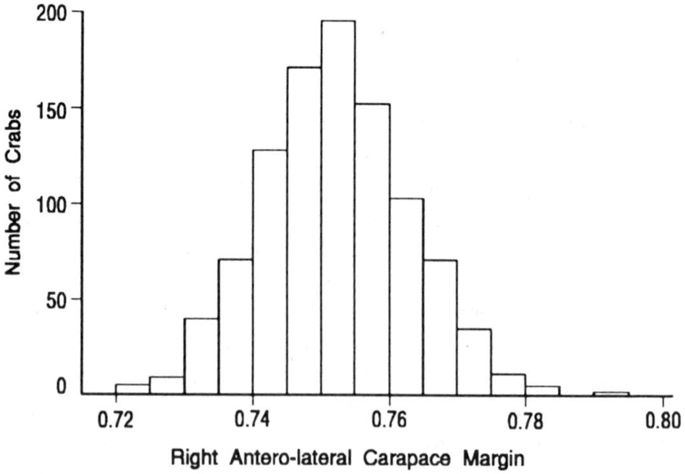

Figure 1.1. *Histogram of the 999 crab carapace measurements listed in Table 1.1.*

fact, he went so far as to assert that there was close agreement between the estimated density and the normal density function.

Figure 1.2 is a more modern density estimate of the crab data that was constructed using methods described in Chapters 3 and 4. Even at a glance it is easy to see the increased resolution in the modern estimate as compared to the histogram. This added resolution makes it possible to double-check Weldon's conclusions regarding the shape of the unknown distribution. To some degree, Weldon's conjecture that the true distribution is normal is supported by the undeniable bell shape of the estimated curve shown in Figure 1.2. However, the curve is not perfectly symmetric, but rather seems slightly wider or heavier on the right side.

This example shows how valuable an estimate of the density function can be as a first step in a data analysis. Before applying many statistical procedures, the researcher is often required to make certain assumptions about the distribution of the data, for instance, that they follow a normal distribution. An estimate of the probability density function affords a convenient and accurate means to check the validity of many distributional assumptions. Yet, although the density function has been the main focus of much of the curve estimation literature, and garners a significant amount of attention in this book as well, it is not always the most appropriate curve to estimate in every circumstance. Take, for

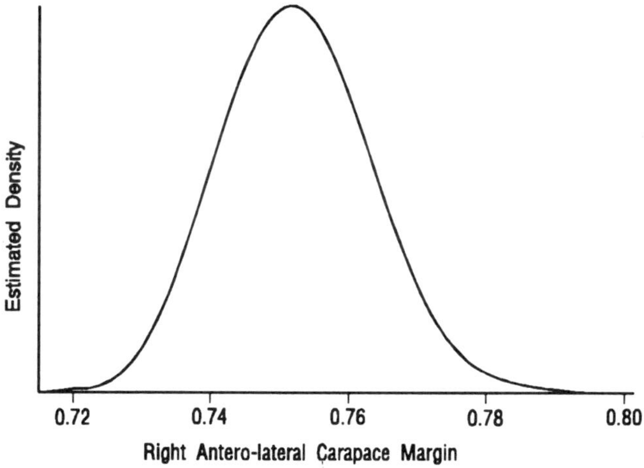

Figure 1.2. *Modern density estimate of the crab carapace measurements listed in Table 1.1.*

example, Weldon's investigation. Weldon wanted to characterize the distributions of each of his measurements, since he considered the shapes of those distributions to be important indicators of the ongoing process of natural selection. He reasoned that if a population was stable, and not actively evolving, then it would contain as many larger than average individuals as smaller than average members. In fact, Weldon believed that in the absence of strong natural selection nearly any characteristic of a species would be normally distributed.

As shown above, an estimate of the probability density function provides one way to assess the normality of a distribution. However, it still requires some subjective input on the part of the investigator. In particular, a fair amount of experience and judgment are needed to gauge just how bell-shaped a curve must be before it can be considered normal. Alternatively, a estimator known as the $R(x)$ curve could also be used for this purpose. As described in Chapters 6 and 7, $R(x)$ is a graphical tool that relies more on human intuition than on statistical experience when evaluating how well a statistical model fits the data. Like the Q-Q plot, the straightness of the $R(x)$ curve indicates how well the model fits, that is, the straighter the line, the better the fit.

Figure 1.3 shows a density estimated from Weldon's data along

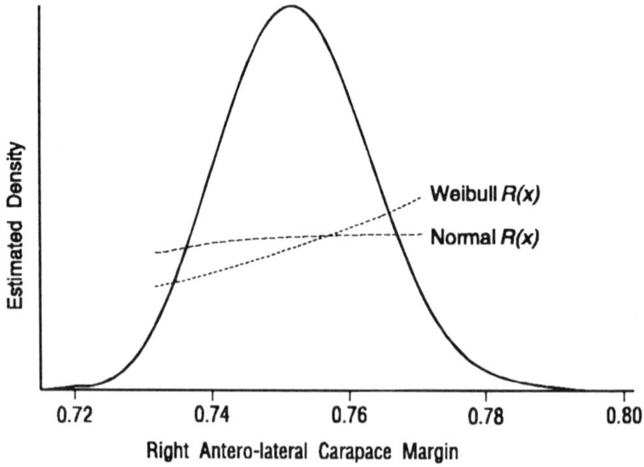

Figure 1.3. *$R(x)$ curves superimposed on the density estimate shown in Figure 1.2. The curve drawn with short dashes corresponds to the Weibull model, while long-dashed curve corresponds to the normal model.*

with $R(x)$ curves corresponding to two different statistical models. The straightness of the long-dashed line, for the normal and the short-dashed line, for the Weibull, indicates how well these respective models fit the estimated density. The slight bend along the left extent of the normal $R(x)$ curve indicates that Weldon may not have been justified in assuming that the data were normal, rather than, for example, distributed with a Weibull density. This conclusion is also supported by our earlier observation that the estimated density did not appear to be perfectly symmetrical. The straightness of the Weibull $R(x)$ curve predicts that this model would fit the data quite well. In Figure 1.4, a Weibull density curve is superimposed on the estimated density, and the close agreement between the two curves suggests that the data are as likely to be Weibull, as normally, distributed.

1.2 Example: Mixture decomposition

Without the benefit of graphical tools like the $R(x)$ curve, Weldon relied on less sensitive displays like histograms to reach the conclusion that of the 22 characteristics he measured (11 measurements each on the two samples of crabs), 21 were normally distributed. One set of measure-

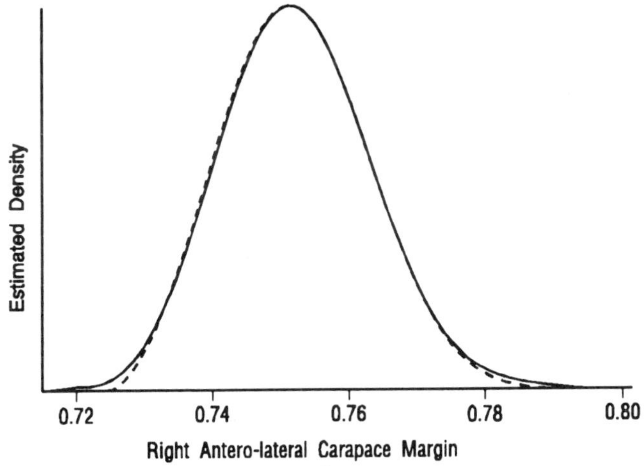

Figure 1.4. *Weibull density (dashed curve) fitted to estimated crab carapace density estimate shown in Figure 1.2.*

ments, however, a feature of the carapace he called the frontal breadth, was distinctly asymmetrical in the sample of crabs taken from Naples Bay. This can be easily seen in the histogram of the data shown in Figure 1.5, and is equally obvious from the density estimate shown in Figure 1.6. Weldon realized that the asymmetry of the distribution could be explained if the sample of crabs had actually contained individuals from two distinct subpopulations, each with its own measurement for frontal breadth. Within each subgroup the measurements would be normally distributed, but the two groups would have different means. If the means were sufficiently far apart, then one would expect to see a density estimate resembling two normal curves side by side. But if the means were close together, the overlap between the two groups would tend to obscure the distinction between them and the result would be an asymmetric density like that pictured in Figure 1.6. Thus, Weldon speculated that the asymmetry of this distribution might be a signal that this population was evolving toward two new subspecies.

Sensing that his own mathematical training was inadequate to confirm this speculation, Weldon turned to his colleague Karl Pearson for help. Pearson's approach assumed that there were indeed two subgroups and that each was normally distributed. He then set about estimating the

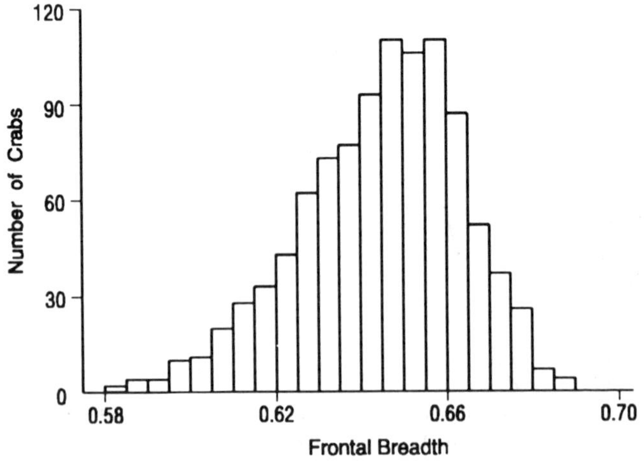

Figure 1.5. *Histogram of 1000 measurements of frontal breadth of female crabs from Naples. The data were collected by Weldon and reported in Pearson (1894).*

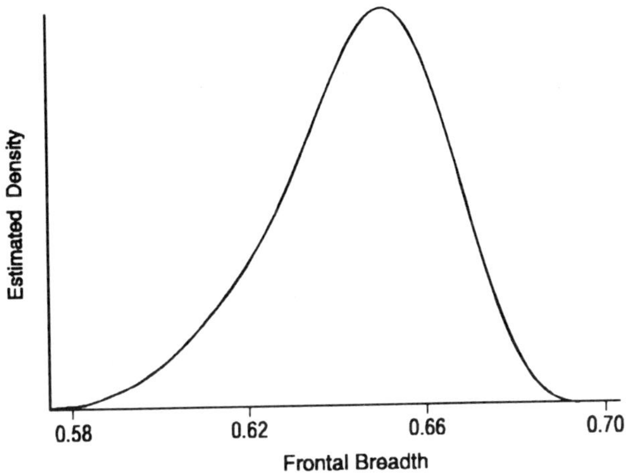

Figure 1.6. *Modern density estimate of the crab frontal breadth data described by the histogram in Figure 1.5.*

means and standard deviations of each group. His paper on this method, published in 1894, was the first to address the statistical problem of mixture decomposition, that is, the decomposition of a distribution into its constituent components.

A curve estimation approach to the mixture decomposition problem, called the λ method, is described in detail in Chapter 5. The λ method differs from Pearson's procedure in that it is mainly a graphical approach, and, more important, because it is not nearly as restrictive in its requirements. Specifically, it is not necessary to know how many subgroups there are in a population to use the λ method, nor is it necessary that any of the groups be normally distributed.

A λ curve estimated from the crab frontal breadth data is displayed in Figure 1.7 along with a rescaled version of the estimated density of the data. Loosely speaking, a λ curve is derived from the original density estimate by reducing the standard deviations of any constituent subgroups without affecting any of the other characteristics of the distribution. Thus, if a distribution is comprised of different subgroups, the λ method enhances the differences between the groups making them easier to see. In this way the method resembles increasing the contrast setting on a television set.

The λ curve shown in Figure 1.7 gives a strong indication that the

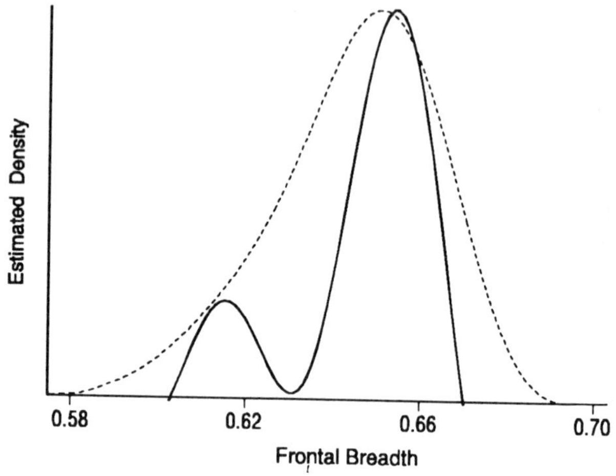

Figure 1.7. *Variance-reduced estimate (solid curve) superimposed on the estimated density (dashed curve) depicted in Figure 1.6.*

frontal breadth data are likely to contain two subgroups, just as Weldon had postulated. Once the overlap between the two groups has been reduced to practically nothing, one of the groups can be eliminated. The λ process that had originally shrunk the standard deviations can be reversed to produce an estimate of the remaining component. The result of applying this procedure to the frontal breadth data can be seen in Figure 1.8, where the right component of the density estimate has been isolated. The shape of the estimate lends support to Weldon's conjecture that the subgroups might be normally distributed. Of course, $R(x)$ curves could be used to investigate this matter further.

The last step in the component isolation procedure represents a subtle point of departure between the curve estimation methods described in this book and other, more traditional forms of data analysis. Both types of approaches begin by estimating some function or curve from a set of data, but the methods differ in how that estimated curve is used. Specifically, most traditional methods use the information from the estimation step to make some decision or judgment about the original data points. Curve estimation methods, on the other hand, frequently derive new types of curves from the estimation step. Thus, rather than return directly to the realm of the data, curve estimation methods often proceed on to new curves which can further the investigation.

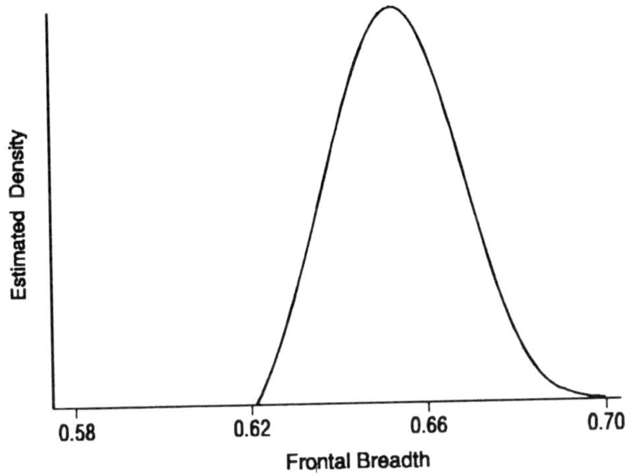

Figure 1.8. *Isolated right-most component of the density estimate shown in Figure 1.6.*

For example, consider the cluster analysis method described in Mardia, Kent and Bibby (1979), which, like Pearson's method, is based on the assumption that the data have been derived from a mixture of normal distributions. Under this assumption, maximum likelihood estimates of the parameters of each component are obtained, and then the estimates are used to determine a decision rule. The decision rule is based on an estimated curve drawn through the range of the data. Points that fall on one side of the curve are deemed to be members of one subgroup, while points on the other side are allocated to a different group. Thus, in this example, a curve is used to make judgments about individual data points.

The λ method, on the other hand, treats the data as a means and not an end, and by doing so, it avoids one of the major drawbacks of cluster analysis. Note that when two populations overlap, it is virtually impossible to draw a curve through the data without some points from both groups landing on the same side of the line. This means that regardless of how well the curve is drawn, there are bound to be points that end up being allocated to the wrong group. The λ method, however, is able to sidestep this problem completely by never attempting to pass judgment on any single data point. Instead the process remains entirely within the realm of curves, leading to density estimates of each of the components of the mixture.

1.3 Example: Regression

Up to this point, both examples have consisted of univariate data sets. Of course, most realistic analyses involve more than one variable. Figure 1.9 is a scatter diagram of 1000 bivariate points artificially generated from a mixture of normal distributions. Since the exact distribution of the simulated data is known (and will be described below), the effectiveness of a curve estimation procedure can be judged by seeing how well it can uncover features of the distribution that are not obvious from simply looking at a scatter diagram.

One of the most popular ways to investigate the relationship between two variables is through regression analysis. Superimposed on the scatter diagram shown in Figure 1.9 is one of several different types of regression curves described in detail in Chapter 8. This type of curve, however,

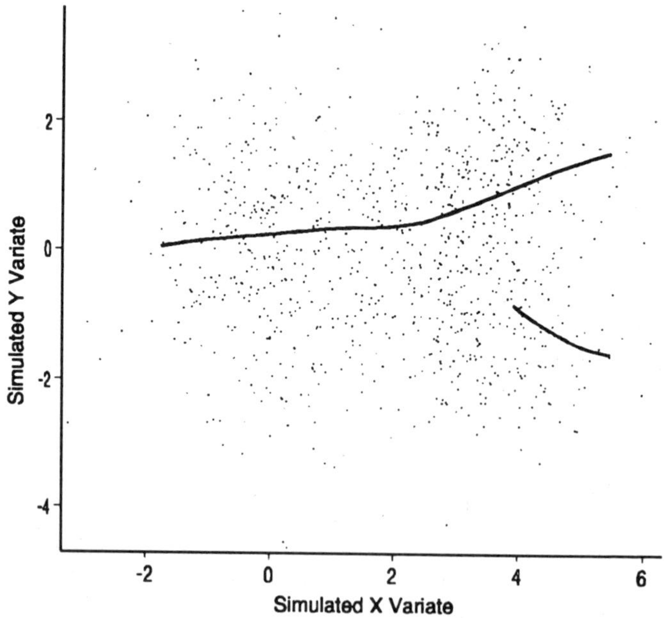

Figure 1.9. *Scatter diagram of 1000 simulated points from a mixture of normal distributions. The curve superimposed on the scatter plot is a mode-based regression which indicates that there are different relationships in different realms of the data.*

is different from any ordinary regression function in at least one important respect: the curve branches into two parts on the right side of the figure. Unlike other regression curves, which are constrained to be single-valued functions, the "mode-based" curve shown in Figure 1.9 is flexible enough to represent multiple relationships between the variables. In this example, the curve suggests that there is a slightly positive, single-valued relationship between the variables throughout the left half of the scatter diagram. But on the right half of the plot, the variables are related in two distinct ways. The two relationships suggest that the right side of the scatter diagram is actually a plot of two groups of data—one with a positive relationship between the variables, the other negatively correlated.

Since the data for this example were artificially generated from a known distribution, it is possible to check on the accuracy of this interpretation of the mode-based regression curve. As speculated above, the right side of the data was indeed derived from a mixture of two bivariate normal distributions with different correlations. The top distribution had a positive correlation of $1/2$, while the bottom had a correlation of $-1/2$. Data on the left side were simulated from a third normal distribution with no correlation between the variables. Thus, in this example, the modal regression curve leads to an accurate estimate of the number of components in the data, and gives a fair indication of the relationships within each group.

Like the λ method described earlier, the first step in constructing a modal regression curve is estimating the underlying density of the data. The only difference is that, in this case, the unknown function is actually a joint density of two variables. A contour plot of the estimated joint density is shown in Figure 1.10. Although it is easy to distinguish the three different distributions in the plot, the nature of the correlation within each group is not as clearly discernible as it was in the modal regression curve.

The next step in constructing the mode-based curve is to compute the conditional density $f(y|x)$ for a series of contiguous values of x. Estimates of these conditional densities can be derived quite easily with only a few simple calculations from the joint density of the variables, which has already been estimated. This process can be thought of as taking cross sections through the function described by Figure 1.10 at several different points along the x axis. At each cross section, the mode or modes of the conditional density are found and their coordinates are plotted. When viewed together, these coordinates form the mode-based

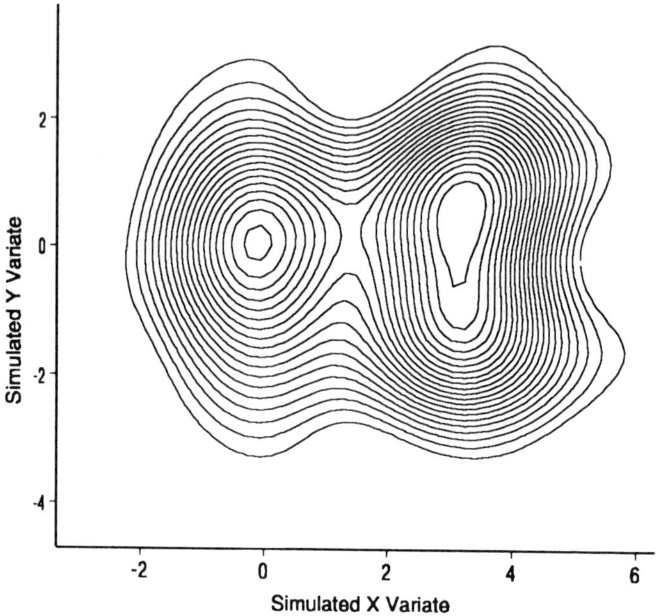

Figure 1.10. *Contours of the estimated joint density function of the data shown in Figure 1.9.*

regression curve. Unlike other properties of a curve, such as the mean or median, a density function can have more than one distinct mode. It is this quality that allows the regression curve shown in Figure 1.9 to break into two parts, technically, to be a multivalued function.

A few of the steps used to construct the mode-based curve shown above are illustrated in Figures 1.11 and 1.12. Figure 1.12 shows three conditional densities taken along the lines through the scatter diagram shown in Figure 1.11. Note that the conditional density taken along Line A has only one mode, and thus, the regression curve is only single-valued as it passes through Line A. In contrast, the density corresponding to a cross section along Line C has two modes, leading to double-valued regression function through the right extreme of the data. The middle of the distribution shares characteristics of both extremes. The curve corresponding to a cross section along Line C shows that although the conditional density at this point consists of two components, there is too much overlap between the components for the density to exhibit

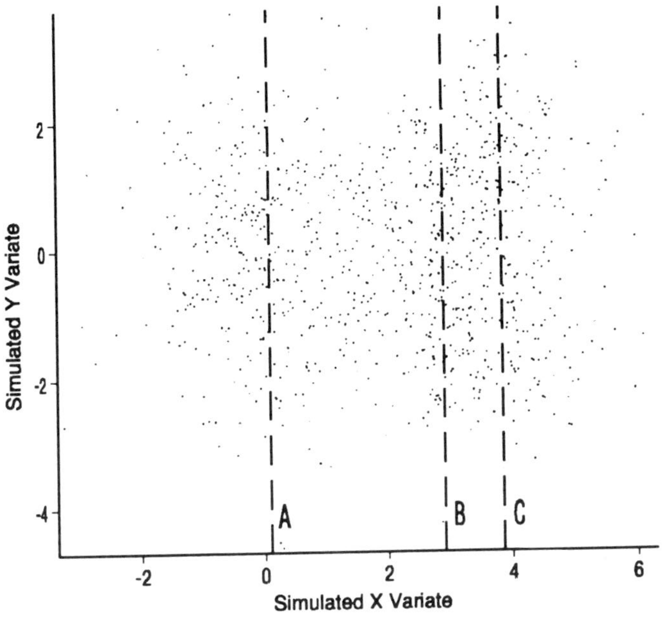

Figure 1.11. *Scatter plot of the mixture data shown in Figure 1.9. The three lines indicate the locations of the three conditional densities shown in Figure 1.12.*

bimodality. Thus, the regression curve takes on only one value as it passes through Line B.

1.4 Outline of topics

This book is divided into three parts. The first part, covered in Chapter 2, is an introduction to the concepts that underlie model-free curve estimators based on generalized forms of representation. The second part of the book consists of two chapters that deal with the basics of pdf or density estimation. Chapter 3 describes basic kernel transform and kernel-based methodology. Chapter 4 deals with the specific details of specialized topics, for example, the application of weighted metrics, which are designed to input a curve user's preference concerning the relative importance of estimator efficiency over different estimator argu-

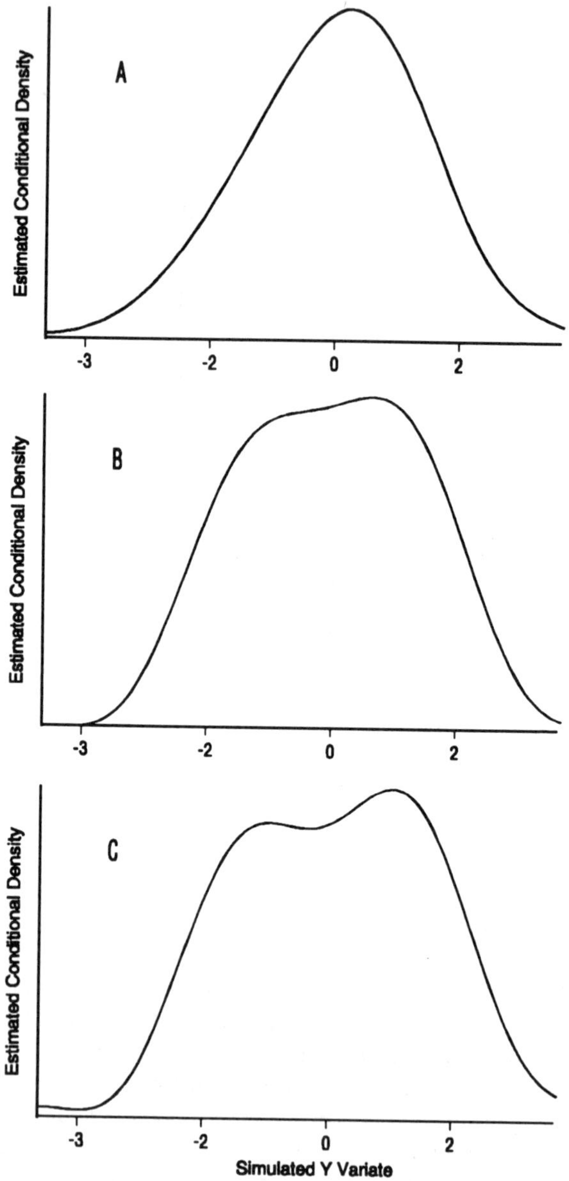

Figure 1.12. *Estimated conditional densities taken along lines labeled A, B and C in Figure 1.11.*

ments. In both of these chapters special attention is paid to the near identity between what are known as kernel-based methods and approaches based on Fourier series representation.

The third part of the book deals with applications and is divided into two subsections. Chapters 5, 6, and 7 discuss intersections between generalized representation and model-based parametric methodology. Specifically, Chapter 5 deals with formulations based on two or more components, like f_1 and f_2, of the classical mixture model, $\{pf_1([x-\mu_1]/\sigma)+(1-p)f_2([x-\mu_2]/\sigma)\}/\sigma$, while Chapters 6 and 7 are concerned with the cumulative (cdf) model G and the transformation v components of the expression $G\{(\mathrm{v}[x-\mu_1]-\mu_2)/\sigma\}$, where, for example, G could be a normal cdf and v could be a logarithmic transformation. While Chapters 5, 6, and 7 deal with one application apiece, Chapters 8, 9, and 10 each deal with a wide variety of applications. Those applications that are connected with maximum likelihood estimation are discussed in Chapters 8 and 9, while Chapter 10 concentrates on applications to nonparametric and other forms of inference.

While the procedures described in this book were all designed to obtain curves that convey information relevant to the population from which data are sampled, there is an aspect of this process that may be new to some readers. This is the differentiation between curve choice and curve estimation. In applications such as those described in Chapters 6 and 7, the two processes, curve selection and curve estimation, are based on a single set of data. Put generally, both overall curve structure is determined and the choice from among alternatives with that structure is made using the same data set. In other applications curve choice, as opposed to curve estimation, is not dependent upon any examination of the data. For example, even today, in most instances where a histogram with a predetermined set of class intervals is constructed, the implicit choice has been made that a particular type of density function (pdf) rather than another type of density, a cumulative distribution function, cdf, or some other type of curve, conveys information relevant to the population being examined.

Simultaneous consideration of curve choice and curve estimation can be subsumed under the heading of statistical generality. Generality is the central theme of this book. In particular, the dichotomization of statistics into problems of specification and problems of estimation, which was first formally proposed by the great statistician R.A. Fisher, is opposite to the generality stressed below.

The processes discussed in this book often involve the construction of a sequence of curves where the observed properties of one curve suggest the choice of the next sequence member. For the multiple steps of statistical exploration, flexibility of curve choice, in the sense of computational convenience, can be as important as data usage efficiency. Some inefficient yet easy to compute estimate of the most definitive curve might, for example, be preferable to an accurate estimator of some curve that is less definitive. Hence, the importance of computational simplicity. Good and Gaskins (1980) state that the particular mode of expression described in this book "has the advantage of involving simpler equations" than other procedures.

There is a proviso to the above that was very succinctly expressed by Tukey (1974) in a commentary on several of the procedures described in this book. According to Tukey, "Fourier expansion is, as we have just seen, an elegant way to do translation invariant smoothing, but it has its peculiarities." We will see that the peculiarities of certain generalized forms of representation can often provide clues to the existence of unsuspected conceptual hurdles and that the surmounting or circumventing of these hurdles can lead to insights of great practical value. Many of these insights are related to the estimation of curves that are more complicated than the pdf.

CHAPTER 2

Generalized representation

By definition, exploratory curve estimation techniques are designed to uncover information directly from data with as few as assumptions as possible. The techniques used must be flexible so that they will yield curves that can conform to almost any shape. To provide this flexibility, all modern curve estimation methods rely on some form of generalized mathematical representation such as Fourier series or kernels. In this chapter the concept of generalized representation is defined and its importance to the field of curve estimation is discussed. The development of generalized representation is traced from Taylor-Maclaurin expansions through Fourier series, and it is shown how some of these representations can be used for curve estimation purposes.

The material presented in this chapter can be classified mostly as background information, and thus, may already be familiar to some readers. Those readers with an extensive mathematical background may wish to skip this chapter and continue with Chapter 3, which begins the exposition on the Fourier series approach to curve estimation. Those who do read this chapter will find it presented at a level made purposefully less technical than the level at which most of the rest of this book is presented. This is because statisticians who make exclusive use of what will be defined below as elementary function models may have grown a little rusty in spots. For example, in regard to the definition of two important quantities, we will show that expressions for the normal and other densities are usually defined in a manner that glosses over a connection between two entities. This connection is central to many topics covered in this book.

2.1 The problem of specification

In their lengthy and beautifully written description of the historical background of nonparametric probability density estimation, Tapia and

Thompson (1978) describe a basic dichotomization of statistical practice where the "problem of estimation" is distinguished from the "problem of specification." While every practicing statistician is familiar with the process of estimation, the term specification—although important enough to be featured prominently in Tapia and Thompson's monograph—is used much less frequently. Stated loosely, specification is finding out enough about a distribution to be assured that certain statistical machinery is applicable. For example, in order to compare two samples by means of a *t*-test, the random variables involved are usually assumed to be normally distributed with the same standard deviation. Thus, by electing to use a *t*-test, the statistician has implicitly assumed that this normal distribution can adequately represent the unknown distribution of the data; that is, he or she has specified that the normal model is adequate for curve representational purposes.

For R.A. Fisher and many other pioneer statisticians, "the problem of specification was to be solved by extra-mathematical means." Tapia and Thompson go on to quote Fisher:

> as regards problems of specification, these are entirely a matter for the practical statistician, for those cases where the qualitative nature of the hypothetical population is known do not involve any problems of this type. In other cases we may know by experience what forms are likely to be suitable, and the adequacy of our choice may be tested a posteriori. We must confine ourselves to those forms which we know how to handle, or for which any tables which may be necessary have been constructed.

This last sentence, outlining the restrictions he placed on the specification process, suggests that Fisher used the word forms to refer to what are today called "elementary functions," specifically, x^k, e^{kx}, $\sin(kx)$, other trigonometric functions, and the inverses and composites formed from x^k, e^{kx}, etc. For example, the normal density function

$$\phi(x) = \frac{1}{\sqrt{2\pi}} e^{-\frac{1}{2}x^2}$$

is composed of elementary functions like the square root and x^2 and, therefore, is an elementary function itself. Almost all of the well-known statistical models, such as the lognormal, Weibull, and logistic density models, can be considered as members of this class of functions.

In contrast, most curve estimation procedures are based on forms of representation that are unlike the expressions for models such as the normal and Weibull densities. A generalized representation is an expression that uses one or more of a variety of notational devices (for example, the series of dots at the end of $k = 0, 1, \ldots$) to escape the typographical and conceptual limit of expression by means of a finite number of explicitly presented symbols. For example, when the expression used to represent a polynomial only contains a finite number of symbols, such representation is not generalized. We will occasionally use the term closed-form to designate nongeneralized representation. Note, according to Noether (1984) the first actual application of the word nonparametric was in Wolfowitz (1942), which contrasted this new branch of statistical inquiry to modes of thinking that involved "one or more parameters, finite in number."

One type of generalized representation employed frequently throughout the book is based on the infinite sequence of functions:

$$\Psi_k(x) = \exp[2\pi ikx], \quad k = 0,-1,1,-2,2,\ldots, \tag{2.1.1}$$

where $i = \sqrt{-1}$. Note that these functions are all orthogonal to each other over the interval [0,1]. Using this sequence, many densities and other functions can be expressed as the Fourier series:

$$f(x) = \sum B_k \Psi_k(x),$$

where $\{B_k\}$ is the set of Fourier coefficients of f (Cencov 1962). Here, the limits of the summation are unspecified and may extend to infinity. It is this lack of restrictiveness which makes Fourier series a form of generalized representation.

Series-kernel and spline techniques, both of which employ generalized representation, have substantial advantages over techniques that use closed-form modes of expression. (Readers who have a particular interest in spline methodology should review Wahba's (1991) definitive text.) Generalized expressions almost invariably have the capacity to represent a very large class of curves, curves that can only be approximated by a closed-form composite containing a finite number of basic elementary functions. The practical value of this approach can be illustrated as follows: Suppose a researcher were to base expressions of the curves or functions encountered in a scientific study, for example, a dose-response curve or hazard function, on one of the generalized representations described in this book. Suppose a second investigator were to base his or her analysis on a more traditional statistical model such as the normal or lognormal. With sufficient data, the first researcher will eventually get the right answer to any properly posed statistical problem. The researcher using the model-based approach could also arrive at the right answer. But no matter how much data were available, he or she could still be misled by the initial choice of the wrong model.

In today's curve estimation literature, there are almost as many forms of generalized representation as there are alternatives to the choice of the normal model within the classical literature. For example, in Good and Gaskins (1980), both the trigonometric and the Hermite system of orthogonal functions were extensively applied to the important problem referred to by Good and Gaskins as series-based "bump-hunting." Anderson (1969), Hall (1980, 1982), and Diggle and Hall (1986) have considered various aspects of estimation based on many different orthogonal systems including the Laguerre, Legendre and cosine functions. The fields of spline, as well as of kernel-based curve estimation, also deal with a large spectrum of choices for the kernel function K or spline representation.

For this diverse spectrum of choices, why is the selection of one particular curve estimation approach different than the selection of a single model in the conventional model-based approach? The answer to this question can be stated in one word: "complete." The importance of completeness is illustrated by the following theorem (we will only define those terms of this theorem that are central to the theme of this introductory chapter):

Given a complete orthonormal sequence in L_2, every element $f(x)$ of the space L_2 admits an expansion, convergent in the mean (Sz. Nagy, 1965). That a curve is L_p implies in practice that the area under the p-th power of the curve is finite and well-defined. In this chapter, it is sufficient to say that L_2 includes at least 99.9% of the curves any practicing statistician is likely to encounter. Since the trigonometric system of functions $\{\Psi_k(x)\}$ that will be discussed in detail in Chapter 3 forms a complete orthonormal sequence for L_2, any L_2 function can be represented by an expansion based on $\{\Psi_k(x)\}$.

Thus, completeness is generality. This does not mean that all methods, all choices of orthogonal function, kernel, spline or metric, will yield identical estimators. Specifically, from a statistical point of view there can be differences in data-use-efficiency and bias. Nevertheless, the mathematical problem of representation is concerned with the question: Can a curve correspond to a given expression? Thus, unlike the tack taken by model-based approaches that rely on elementary function representation, no statistical quest is ever doomed to failure from its start simply for representational reasons. Completeness implies that any function or curve can be adequately represented.

Lack of completeness leads to the problems of specification referred to by Tapia and Thompson. Such problems involve the following two basic questions:

(1) How likely is it that the curve a scientist is actually estimating can be expressed as an elementary function?
(2) Given that the curve can be expressed as an elementary function, how likely is it that the scientist will know what particular elementary function represents the curve?

For example, a mathematical proof that the normal inverse cumulative, Φ^{-1}, could not be expressed as an elementary function was presented by Rosenlicht (1975) as a consequence of a theorem of Ritt (1948). In contrast to the normal inverse cumulative, the completeness of an orthogonal system like $\{\Psi_k(x)\}$ guarantees that a form of representation will be adequate for any given L_p curve, where p is any number greater than 1 (Carleson, 1966).

One of the principal advantages of the methodology described in this book is also a major drawback. The term generality can give the false impression that the field of curve estimation contains ethereal subject matter and that an introduction to its mysteries requires that the statistician or data analyst first become conversant with celestial mathematical concepts such as Hilbert spaces and martingales. Far from being transported to celestial regions, most statisticians who use a histogram, life-table or Q-Q plot are unaware that they are actually using curve estimators. While older tried-but-true methods are often treated as work-a-day tools, modern extensions of histograms, life-tables, fractile diagrams (Hald, 1952), and scatter diagrams are commonly viewed from the opposite and equally distorted perspective of being esoteric gadgets or curiosities that the busy applied data analyst with both feet on the ground can postpone learning about.

It is hardly surprising that mathematically oriented statisticians should be interested in curve estimation. The trend toward generality in statistics has many parallels to earlier fruitful trends in mathematics. These parallels, together with the subtle questions that curve estimation has a tendency to uncover, have led to a very great number of research papers that deal with theoretical aspects of this topic. Unfortunately, many of these papers are written at a level of abstraction that renders their contents inaccessible to all but a few practicing data analysts.

The abstraction of much of the current curve estimation literature can have serious consequences for statistical practice when it acts as a barrier that prevents generalized approaches from being considered as alternatives to their model-based counterparts. It is a major thesis of this book that the field of statistics has been dominated by attention paid to elementary functions for far too long. There is an almost mythical quality of e^{kx}, $\sin(kx)$, x^k and their inverses and composites that reminds one of the twelve Olympians of Greek mythology. Each basic elementary function, like each Olympian, is often treated as having its own character. For example, because it has historically been the forebear of other basic elementary functions, x^k assumes the elementary function mantle of Zeus, while certain inverse functions play the role of the tricky and mercurial Hermes.

It is not surprising that when basic elementary functions are personified, identities are also given to location and scale parameters as well as properties such as the mean, mode, median, standard deviation, skewness and kurtosis. In an important sense, parametrization almost always is tantamount to personification. On the other hand, representational completeness is the one common element of the nonparametric viewpoint. Whether there are twelve individualized entities or twelve hundred is only a matter of statistical and computational efficiency.

2.2 History of generalized representation

The early development of generalized representation was certainly not a logical progression of interconnected and related ideas. For this reason, this section provides an idealized overview and not a historical summary of past and recent developments. As discussed in the preceding section, modern forms of curve estimation are all based on generalized representation which contrasts starkly with the closed-form expression of what below will be called "a model." Specifically, models are composed of only a fixed finite number of basic elementary functions, and thus, do not have the adaptive quality of generalized representation.

An important characteristic of generalized forms of representation is introduced by considering the role played by the constant 1/2 within the exponent of the normal density model

$$f(x) = \phi[(x-\mu)/\sigma]/\sigma = (\sigma\sqrt{2\pi})^{-1} e^{-\frac{1}{2}[(x-\mu)/\sigma]^2}.$$

The primary purpose served by this constant is to assure that the value of the scale parameter σ within the normal density model equals the property most commonly connected with scale, the standard deviation:

$$\sigma = \left[\int_{-\infty}^{\infty} (x-\mu)^2 f(x)dx\right]^{1/2}.$$

Ever since Descartes introduced the word parameter as that part of the analytic geometry expression for the property of a parabola which corresponds to the specific property—diameter of a circle; parameters, have been associated with properties. Until we more fully define the two terms, parameter and property, in Chapter 6, suffice it to say that a parameter is a symbolic component within either model-based or generalized representation; while a property, like the standard deviation, is a number that characterizes a curve.

A property like the standard deviation characterizes a density in much in the same way that the determinant of a matrix characterizes a matrix. Yet in a crucial respect this is a bad analogy. While the computation of a matrix's determinant is central to many applied and theoretical aspects of matrix algebra, little consideration has been given to the question: how can some single element of a matrix be linked to the determinant? On the other hand, a single parameter of a curve's representation is often linked to one or more properties like the mean or standard deviation. For example, the location parameter μ of $f(x)=\phi[(x-\mu)/\sigma]/\sigma$ of is commonly linked to the three distinct properties, mean, median and mode.

That the scale parameter σ is commonly linked with only one property but the location parameter μ is sometimes linked to three properties illustrates a distinction between elementary function-based models and the Fourier series representation $f(x) = \Sigma B_k \Psi_k(x)$. Within this later expression B_k, for a fixed value of k, serves a role similar to that served by either μ or σ within the expression for the normal model. An important difference between the Fourier series form of generalized representation and a typical elementary function, is illustrated by the important curve property, $\int_0^1 f(x)e^{-2\pi ikx}dx$. This property is so closely connected with the parameter B_k that it is called "*the* Fourier coefficient." Loosely stated, there is as little ambiguity concerning the connection between properties and parameters in this case, as there is in the designation *The* Bronx.

Clearly, the connection between (1) parameters and their generalized equivalents, for example, Fourier coefficients and (2) properties is central to the representational process. In the Fourier series and in many other instances, this process is called "expansion." The first form of generalized representation, the Taylor-Maclaurin (TM) expansion, was devised by Brook Taylor as a modification of a method used by Newton and other seventeenth and eighteenth century mathematicians to interpolate table values (Kline, 1972, pp.440–442). To facilitate comparison with other representational techniques, this process is illustrated by the circle drawn above the point x_0 on the x-axis in Figure 2.1, which shows an idealized magnifying lens used to visualize locally a sequence of curve properties. This sequence consists of the evaluation of some function h, as well as evaluations of all of the derivatives of h at x_0, through which the TM expansion process assembles a representation of h by using local properties of h at x_0 to join together a weighted linear combination of the polynomials $(x-x_0)^k$, $k = 0, 1, \ldots$. Whenever it is

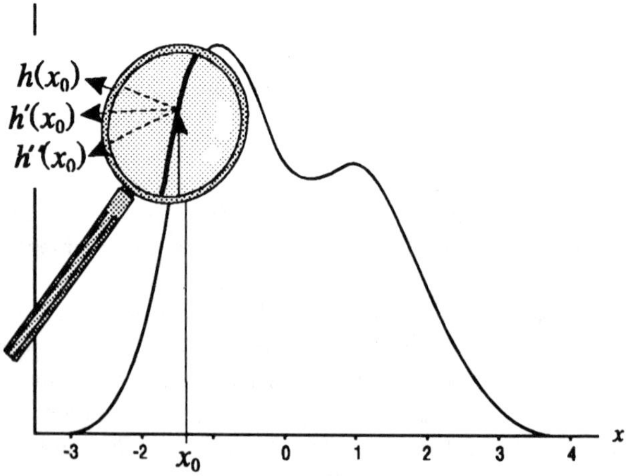

Figure 2.1. *TM expansion process. The magnifying glass emphasizes that a TM expansion is based on the local properties of a curve at one specific point.*

useful to stress that it is not necessarily a density that is expanded, we will use the symbol h and not f, the letter commonly used to represent both a general function and the density.

After the TM expansion process the next logical step leading to today's forms of generalized representation can be conceptualized as being the replacement of the sequence of curve properties at a single point with an sequence of equally spaced evaluations of h (Bernstein, 1912). Like the TM expansion, the Bernstein expansion is a weighted combination of the polynomials; specifically,

$$\frac{n!}{k!(n-k)!}x^k(1-x)^{n-k},\ k=0,1,\ldots,n.$$

Lorentz (1953) and Rosenblatt (1962) give clearly written descriptions of the reasoning underlying this form of expansion.

As illustrated in Figure 2.2, and unlike the TM process, the component polynomials used to form the composite polynomial of the truncated Bernstein expansion are linked by the global, equally spaced evaluations of h. Both the TM and the Bernstein expansion processes are based directly on properties of the function being expanded. However, curves can have dual identities. Hence, the question: Could the properties

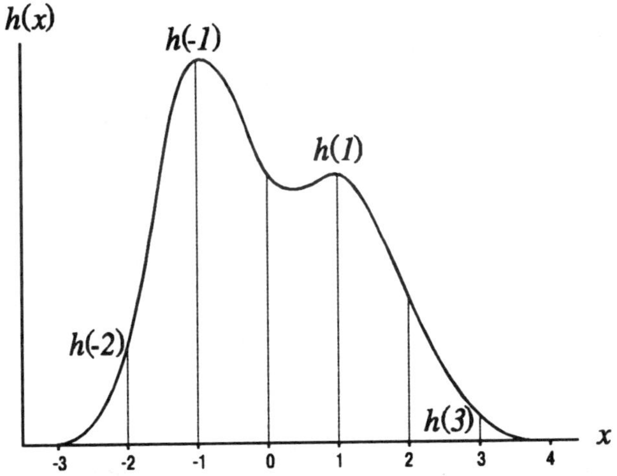

Figure 2.2. *Bernstein global expansion process. A Bernstein expansion is based on equally spaced evaluations of a curve.*

of some dual of a function be useful for representational purposes? Historically, the first attempt made by statisticians to answer this question was based on the curve properties, ordinary moments about the origin, $\mu'_1, \mu'_2, \ldots$. When $\mu'_1, \mu'_2, \ldots$ are used as part of the representation of a function h, this can be viewed as expression in terms of properties of a second, dual, function, the moment generating function (MGF) whose properties are evaluated in the same way that the TM expansion evaluates the properties of h itself—as evaluations of derivatives at a single point. Hence, when moments exist, they are local evaluations of the dual function.

It is important to note that the function h itself is not the actual curve used to connect representation subcomponents. Instead the curve is an alias or dual form of h. The basic idea that functions can be paired or interconnected is central to Newton's and Leibnitz's 17th Century Calculus. We will reserve the word transformation for processes that are to random variables what a transform is to curves or functions: specifically, the pairing or association of entities. While a transformation can be often treated as simply being some function itself, a transform usually should be considered to be a function of a function. The transform of a function is a function only in the same way that the value assumed by a function is a real or other type of number. Thus, when a name is given to a transform, as in "Fourier Transform," this name refers to a type or class and not to a particular function, like the normal density. However, a normal density might be a Fourier transform of some other density or even the transform of the normal density itself.

Examples of transforms include the MGF and the characteristic function (CF) or Fourier transform. The CF of a density $f(x)$ can be written as

$$\psi(t) = \int_{-\infty}^{\infty} \exp(-2\pi itx) f(x)\,dx, \tag{2.2.1}$$

which is equal to

$$\psi(t) = \int_{0}^{1} \exp[-2\pi itF^{-1}(y)]\,dy \tag{2.2.2}$$

for any distribution with a one-to-one cumulative F and inverse $F^{-1}(x)$. (The MGF $\eta(z)$ can be defined as $\psi[iz/(2\pi)]$.) The process of determining $\psi(t)$ using knowledge of f or $F^{-1}(x)$ is of considerable importance to applied mathematical scientists and is the subject of an immense literature of computer software and hardware, articles, tables, and books (for example, Ramirez (1985)). Because the word transform is much more widely used than the words dual or alias in reference to curves or functions, in all later chapters we will use it to refer to the CF.

When a density f has a MGF η, or a CF ψ, η or ψ are no less representations of the underlying distribution than is f. η is called the MGF *of f* only because educational experience familiarizes us with a density like f before we are introduced to the MGF function. Were our schooling different, it would seem as natural for the properties of the MGF of f, rather than the properties of f, to be used as the starting point from which a distribution's curves would be expanded. As sketched in Figure 2.3, either the MGF or the CF can be represented as a TM expansion about the point $x_0=0$. Efforts by Gram, Charlier, Edgeworth (GCE) and Pearson to estimate curves based on evaluations of the MGF and its derivatives about $x_0=0$ were, on the whole, unsuccessful (Boas 1949; Hildbrandt, 1931; Samuelson, 1943). The ordinary moments from which the MGF and CF can be represented as TM expansions did not provide representations suitable for many statistical applications.

The process that underlies the Fourier expansion of a pdf, f, or a cdf, F, sketched in Figure 2.4 is much more straightforward than is the GCE expansion process. Here, the CF ψ, or the closely related Fourier transform, serves as the dual or alias function used to link together components of a generalized form of representation. However, while the GCE and Pearson family procedures use the ordinary moments, which are *local* CF properties, Fourier expansion, like the Bernstein expansion process, is based on an equally spaced sequence of curve

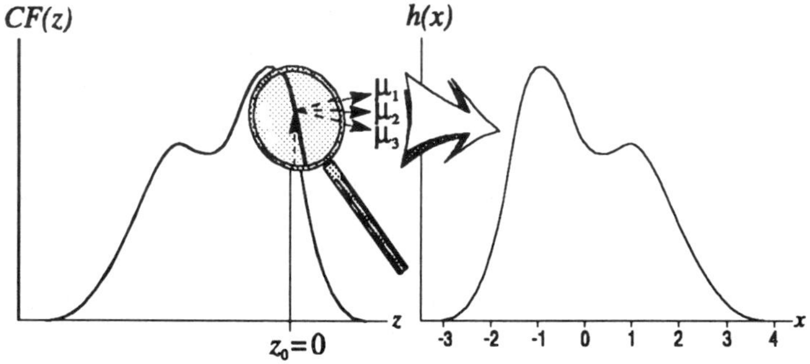

Figure 2.3. *Moment-based expansion showing use of alias CF function locally evaluated about the single point zero.*

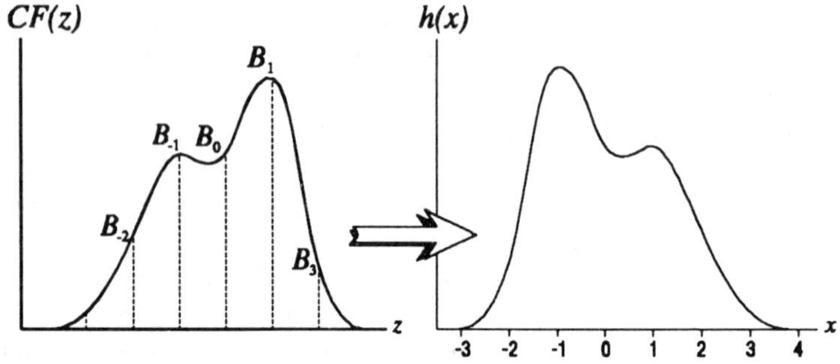

Figure 2.4. *Fourier-type expansion showing use of alias CF function globally evaluated at a sequence of equally spaced points.*

evaluations. But, like ordinary-moment-based methods, and unlike the Bernstein process, it is the properties of a transform of the function being expanded, the CF, and not equally spaced evaluations of f itself, that provide the linkage of f's representational components.

2.3 From representation to curve estimation

Up to this point, we have considered the issue of representation from a purely mathematical point of view. But for the statistician, representation is far less of an end in itself than it is for the mathematician. According to the historian of mathematics, M. Kline (1953, p. 377):

> Arguing from the existence of the mathematical laws uncovered during the Newtonian era, the eighteenth-century thinkers erected the most comprehensive and influential philosophical system of modern times. It propounded a world that is designed and orderly and that functions according to plan. Mathematical laws made manifest that design, and the unfailing fulfillment of scientific predictions gave proof that the design was being adhered to. . . .
>
> The philosophy of determinism still dominates our thoughts and beliefs and guides our actions. Unfortunately, the order of nature supremely simple and harmonious to the founders of modern science, now appears to be breaking up in the maelstrom of statistics and probabilities which the nineteenth and twentieth centuries have used so effectively.

The limitations of determinism in science have focused a great deal of attention on the bifurcation of deductive logic and the realistically complex inductive processes of statistics. In curve estimation applications the function to be represented is never known. Thus, an additional step must be added to the expansion process. This step can be interpreted as the union of the problem of specification and the problem of estimation. For the Pearson family and GCE approaches, it involved the substitution of entities that were called "sample moments" and "k-statistics", for their population counterparts, a process which can be called the method of sample moments. What sample moments and k-statistics actually are, is of less importance than that there is a linguistic pitfall. The naming of an estimator as the "sample-something" suggests, but does not in any way verify, that the sample-something is the most natural and useful choice of estimator. From the perspective of today's massive statistical literature, there may be many alternatives to any statistic. For example, even though the standard deviation of the normal is the single property which is almost exclusively associated with the scale parameter of this model there does not exist any estimator the normal density's scale parameter which is optimal in all respects.

Like the method of moments, many topics considered in this book involve the replacement of Fourier series sample somethings for their population counterparts. For example, a sample equivalent of the CF defined by (2.2.2) can be constructed from an iid sample of data, $\{X_i\}=X_1,\ldots,X_n$ from the density f:

$$\hat{\psi}(t)=\int_0^1 \exp\left[-2\pi it\hat{F}^{-1}(y)\right]dy \tag{2.3.1}$$

where for $j=1,\ldots,n$, the sample inverse cumulative $\hat{F}^{-1}(z)$ is defined within each half open interval $((j-1)/n,\ j/n]$ to equal the jth sample order statistic $X_{(j)}$. (By definition, the order statistics satisfy $X_{(1)} \leq X_{(j)} \leq X_{(k)} \leq X_{(n)}$ if $1 \leq j \leq k \leq n$.) Expansion based on the empirical, or sample, characteristic function (SCF) is discussed by Blaydon (1967), Brunk (1978), Davis (1975), Feuerverger and Mureika (1977), Kronmal and Tarter (1968), Rossi and Brunk (1988), Tarter and Kronmal (1970, 1976), Tarter, Kronmal and Holcomb (1967), Tarter (1979a,b; 1986), Wahba (1981), Watson (1969), Watson and Leadbetter (1963), and Winter (1979). While this form of expansion is based on the Chebyshev orthogonal polynomials associated with the most commonly applied form of Fourier series, expansion based on the Hermite orthogonal polynomials associated with the normal density is discussed by Schwartz (1967, 1969a,b), Walter (1977), as well as Good and Gaskins (1980). Expansions based on the orthogonal system known as the class of Walsh functions, Walsh (1923), Lackey, (1972), are described by Brunk et al. (1972); Brunk and Pierce (1974); Butler and Kronmal (1985); Fukunaga and Ito (1963); Gose (1965); Hawkins (1961); Moore (1973); Ott (1972); Ott and Kronmal (1976); Tarter (1986); and Uesaka (1971).

Almost all modern curve estimation procedures involve a step that has no method of moments counterpart. Besides the SCF, another curve is involved. Figure 2.5 sketches an underlying approach to curve estimation based on series, kernels or transforms where two alias functions are utilized. For orthogonal function system $\{\Psi_k(x)\}$ representation, the first of these curves is the SCF. The SCF provides a mechanism for incorporating the data into the curve estimation process, and also plays a role similar to the one played by the CF in curve representation. That is, equally spaced evaluations of the SCF can be used to represent an

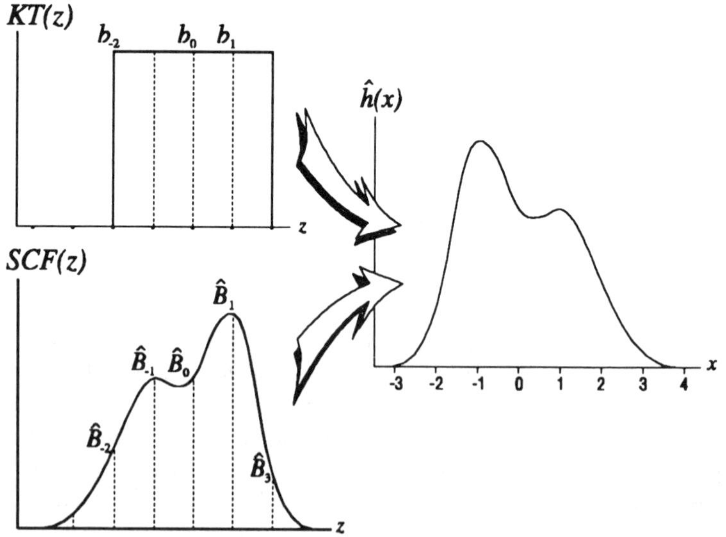

Figure 2.5. *Curve estimation based on Fourier-type expansion showing use of dual alias functions, KT and SCF, each globally evaluated at a sequence of equally spaced points.*

estimated curve in the same way that equally spaced evaluations of the CF can be used to represent a known curve.

The second form of alias curve which, together with the SCF, provides the framework for many procedures described in later sections of this book, is the kernel transform (KT). Here the word kernel designates that the second alias function is the Fourier transform, not of the density itself, but of some function $K(t)$, which can be interpreted in terms of the spreading of zero-dimensional data points over one- (for the case of the univariate generalized histogram), two- (for the case of the generalized scatter diagram), or higher-dimensional spaces.

The KT is necessary because there is no known way to construct statistical machinery that simply takes data in and spews knowledge out. Instead, almost all advances of modern theoretical statistics involve sorting out processes by which utilities, losses, distances, metrics and priors impact data-to-knowledge conversion. Although it does not subsume the term prior, we will use the word "criteria" to designate all other of the above formalizations. The role of the KT is to facilitate the integration of criteria, and other forms of user input, with data.

For many applications, the KT is the vehicle through which information supplemental to the data points themselves is input into the estimation process. An example of this form of information is the relative importance of estimator precision over specific curve support subregions containing the running variable x, as specified by the weight function $w(x)$ within the mean integrated square error (MISE) metric:

$$E\int_{-\infty}^{\infty}\left[\hat{f}(x)-f(x)\right]^2 w(x)dx, \tag{2.3.2}$$

where $\hat{f}$ is an estimate of the curve f. Usually, both the KT itself and its Fourier transform, $K(t)$, have only indirect associations with the purposes of some given statistical application. But, as detailed by Tarter, Freeman and Pollisar (1990), $w(x)$ is directly connected with such problems as the enhancement of long term survival curve estimator accuracy. Hence, the pair of curves, KT and the SCF, serve as the medium through which information in the sampled data and a user selected $w(x)$ impact the curve estimation process.

There are two reasons why the function w was chosen to illustrate the role played by the KT. First, the role of w is easier to conceptualize than the roles played by other inputs that influence the KT. The second reason is that while w weights information in the data domain, the KT weights information in the frequency domain. From a general point of view, series-expansion forms of curve estimation can be based on the mapping of both $\{X_j\}$ and w onto the pair of alias functions SCF and KT. In the actual computation, individual equally spaced evaluations of the KT (called "multipliers") are paired with counterpart individual equally spaced evaluations of the SCF given by (2.3.1).

Implementation of curve estimation procedures based on two rather than one alias function also has the following advantages:

(1) Curve expansion based on the CF and Fourier series is the method chosen for most series or transform applications. Hence, by applying the SCF's simple bias and variance properties it is possible to extend the advantages of Fourier-based expansion to deal with a wide variety of curve estimation problems.

(2) The curve estimation process begins with the specification of the criteria by which the effectiveness of curve estimation alternatives and modifications can be judged. We will see in the next two chapters and in Chapter 9 that the mean integrated square error metric with weight function w is a particularly tractable criterion for judging between alternative series forms of curve estimators. Based on the MISE metric, information from the sample is then used to determine a KT, which in turn yields an associated set of multipliers. Consequently, the SCF can be viewed as the computational and conceptual framework about which the detailed and individualized superstructure of data analysis can be worked and re-worked by different choices of KT.

2.4 The method of moments

As noted at the beginning of the previous section, the representation of functions by transforms or other means is chiefly a mathematical rather than a statistical problem. No matter how a representational process works, it is necessary to input data to estimate the coefficients of any representation. In many applications that use curve estimation methods, at least two curves must be estimated. The association between expansion coefficients and certain properties like moments or equally spaced CF evaluations does not dictate what should be done with data. No better illustration of this crucial point exists in the statistical literature than the estimation of the coefficients of GCE and Pearson family expansions.

On page 147 of his classic text, Kendall (1952) introduces the GCE process by stating that, "The reader familiar with harmonic analysis will recognize the resemblance between the [GCE] procedure and the evaluation of constants in a Fourier series." Surprisingly, despite the similarity of GCE methods and more modern approaches to curve estimation and generalized representation, this connection is seldom, if ever, made. This is in part because GCE analysis, like Pearson family methodology, used the method-of-moments and did not take the crucial step of utilizing a metric such as the MISE.

The non-metrical approach to GCE and Pearson family methodology was largely due to a quirk of history. Karl Pearson developed the Chi-square procedure whose simple extension to tests of goodness-of-fit is closely related to the weighted MISE metric. The GCE expansion process was primarily used by members of what Sarndal (1971) refers to as the Scandinavian School of statistics. However, while the GCE expansion process is tractable with respect to MISE (for example, see Schwartz (1969a, b)), MISE compatibility does not extend to the Pearson family. Hence, the method of moments, essentially a false analogy between entities, was the only approach considered for estimation of the coefficients of both the Pearson and GCE forms of generalized representation prior to the 1960s.

The Pearson family and GCE forms of representation are rarely used today for curve estimation. This is in part because GCE and Pearson family methodology depend on two questionable assumptions. The first is that the representation of a curve in terms of population moments implies that the estimation of that curve should be based on what seem to be the sample equivalent of these population moments, or their close relative, k-statistics. That k-statistics are both difficult to compute and tend to yield poor curve estimates were clues that something was wrong with the method of moments as the basis of a curve estimation technique.

The second assumption that underlies the use of the GCE and Pearson family methods is subtle. It was assumed to be desirable for a generalized form of representation to have the normal as a special case. The normal depends on two parameters commonly estimated by the sample mean and sample variance. This led to the specious conclusion that the multi-term Pearson family and GCE estimates should also be based, at least in part, on the sample mean and sample variance.

When $g_p(z)$ and $g_G(z)$ represent the derivatives of the natural logarithm of $f(z)$, where $f(x-\theta)$ is the density of the jth member X_j of an iid random sample, then the functions

$$g_p(z) = z/[\alpha_1+\alpha_2 z+\alpha_3 z^2] \tag{2.4.1}$$

and

$$g_G(z) = \alpha_1 z+\frac{\alpha_2+2\alpha_3 z}{1+\alpha_2 z+\alpha_3 z^2} \tag{2.4.2}$$

are the quadratic special cases of the Pearson family and GCE expansions, respectively. Both for Pearson family and GCE representation, setting $\alpha_2=\alpha_3=0$ yields the normal special case. Both of these methods were generalized by using higher-degree polynomials in place of the quadratic denominator $\alpha+\alpha_2 z+\alpha_3 z^2$. However, GCE higher-degree polynomial extensions of the denominator were expressed in terms of orthogonal polynomials known as the Hermite system while the Pearson model simply treated such extensions as general polynomials. Pearson's equation also differs from the GCE equation for $g_G(z)$ in having a large number of commonly encountered statistical models besides the normal as special cases. However, like GCE expansion (2.4.2), it neither has the lognormal model nor the most common mixture of densities as a special case.

Pearson family representation has a connection with series types of orthogonal expansion. A basic mathematical relationship, Rodrigues' formula for the generation of orthogonal polynomials, is known to apply to orthogonal polynomials associated with the $g_p(z)$ special case of the Pearson family (Abramowitz and Stegum, 1964, pp. 785; Peebles 1940).

The argument that the Pearson family is useful because it contains many commonly used statistical densities is circular. It is true that Rodrigues' formula applies to many curves applied by statisticians and mathematicians. But the simplicity of Rodrigues' formula, which connects the generation of orthogonal polynomials to the operation of differentiation, has focused much mathematical study on models that happen to be special cases of the Pearson extended model. It is therefore not surprising that these well-studied models were tried as models for statistical densities. However, just because these models satisfy Rodrigues' formula does not imply they are in any sense appropriate for applied problems.

The beginnings of today's revolution in computational capability occurred during the decade in which Elderton's (1953) definitive book on the Pearson family was published. It is small wonder that prior to the 1950s emphasis was placed on the parsimonious mathematical representation of curves. At that time only modest capability was available by which complex forms of curve estimation could be studied empirically. It was not until modern digital computation and curve estimation approaches became available that moment-based Pearson family and GCE representation were found unfruitful. These studies raised doubts concerning the individualization or personalization of parameters within elementary function-based models. This, in turn, led to new approaches in which priors, pilot estimators, curves, penalties and metrics supplemented or replaced parameters.

The processing of data by means of the SCF and the KT differs radically from pre-1956 methods based on the GCE expansion and Pearson family representation. The assumption of a metric rather than the assumption of a model involves identifying what is or is not important to the researcher. It decidedly is *not*, as are Pearson family or GCE representations, a part-way guess, with some blanks left to be filled in amidst specified information about the data generation process. When a symbol for a density like f appears as part of a metric like the MISE, it is usually paired with a symbol like $\hat{f}$, which represents an estimator of f. Loosely stated, the rest of the metric specifies how departures of $\hat{f}$ from f evaluated at one point should be emphasized or de-emphasized relative to departures of $\hat{f}$ from f at some other point. A key concept is that through the choice of w, trade-offs are judged from the subjective point of view of the scientist and not, as in the case of an extended model like the Pearson or GCE g's described above, some partial set of assumptions concerning the state-of-nature.

CHAPTER 3

Series and kernel-based density estimation

In an extremely perceptive article, J. Tukey (1965) picturesquely used the term "Erector set" to refer to what he called the "kit of pieces" that constitutes the practicing statistician's methodology. Along these same lines, this chapter deals with the nuts and bolts of curve estimation, that is, the kit of pieces from which many types of curve estimates can be constructed. For simplicity we will start by concentrating on the estimation of one particular curve: the probability density function. In later sections we will then show how these density estimation techniques can easily be adapted to deal with other curves, such as the cumulative distribution function. Also, for simplicity, we will begin with the univariate case; specifically, suppose we have an independent sample $\{X_j\}\equiv\{X_1,..,X_j,..,X_n\}$ whose constituent variates have common pdf f. The question then arises: How can this unknown density be estimated from such data?

3.1 Initial choices

Every choice of statistical technique is based on initial assumptions. For example, if a variate is assumed to be normally distributed, or more commonly, if it is believed that it can be transformed to normality, a t test is often deemed applicable. A linear model of response and predictor variable relationship is another commonly encountered assumption. Like parametric approaches, semiparametric and even nonparametric methods do not exist in vacuo. For instance, independence of measurements is frequently required. Thus, it is not the lack of assumptions that characterizes general statistical approaches; rather, it is minimal restrictiveness within the confines of necessary assumptions.

At times, justification can be based on theory. For example, under mild conditions, the certainty that the mean of a sample can be taken to be a normal variate increases as the amount of data increases. Another approach to the specification problem is based on study of some data-producing mechanism. Occasionally this mechanism can be so well understood that complete confidence is justified. Finally, although not the least frequently, assumptions are made purely for reasons of convenience. A linear model is often chosen simply because it provides a framework for easily linking together many predictor variables in terms of access to some prepackaged computer software system.

The curve estimation methods described in this chapter also require initial choices, the two most important of which are the mathematical form used to represent targeted curves and the metric used to judge curve accuracy. When the goal is to search for clues regarding distributional structure, techniques must be flexible enough to disclose any subtle or complicated pattern in the data. In such applications it is obvious that any initial choice or assumption must not be restrictive, in the sense of constraining flexibility. The choice of methodologies presented in this chapter has been made with this flexibility very much in mind.

The first choice—the form of representation—has already been introduced in the previous chapter. Until Chapter 5 it will be assumed that estimated curves can be represented by a uniformly convergent Fourier series over the support region [0,1]. Below, the unknown density can be expressed over the unit interval as the uniformly convergent series

$$f(x) = \sum_{k=-\infty}^{\infty} B_k e^{2\pi ikx},$$

where $i = \sqrt{-1}$ and the Fourier coefficients, B_k, k=0,±1,±2,..., are the properties

$$B_k = \int_0^1 f(x)e^{-2\pi ikx}dx.$$

In Chapter 5 a procedure developed by the mathematician Fejér will be shown to allow the above assumption to be safely made for all densities of a continuous random variate. In Chapter 8 procedures described by Lanczos will be shown to allow this assumption to apply to any discrete or binary valued variate. Estimators of the density are obtained by first estimating a subset of the true Fourier coefficients and then constructing a Fourier expansion similar in form to the above expansion of $f(x)$.

Thanks to advances made by Wald (1947) and many others, after a representational medium is selected, a second decision is now taken to be de rigueur. It is deemed vital to assess the accuracy of the estimates themselves, that is, to choose a metric. The metric selected throughout most of this book is one with a natural and historical connection to Fourier series estimators, the weighted Mean Integrated Square Error, or weighted MISE:

$$J\left(f,\hat{f},w\right) = \mathrm{E}\int_0^1 \left\{f(x)-\hat{f}(x)\right\}^2 w(x)dx. \tag{3.1.1}$$

Here $\hat{f}$ is the estimated curve and w is a user-defined weight function. The most frequently considered weight function, and the one which we will begin with as well, is $w(x)\equiv 1$. This leads to the unweighted MISE, or more simply, the MISE:

$$J\left(f,\hat{f},1\right) = \mathrm{E}\int_0^1 \left\{f(x)-\hat{f}(x)\right\}^2 dx. \tag{3.1.2}$$

A good estimate, then, is one which minimizes the MISE. In the next chapter we will present several methods for optimizing estimators designed to reduce either the unweighted or weighted MISE.

The following is an illustration of the historical importance of metrics similar to the MISE in conjunction with Fourier series methods. The classic work that deals with general orthogonal systems was published by Szego (1939). The first chapter of this book deals with criteria and is entitled "General Properties of Orthogonal Polynomials." The first theorem given in this chapter deals with what Szego calls, "the weighted quadratic deviation." With the exception of expected value operator E of the MISE, the weighted quadratic deviation is identical in form to the MISE.

The MISE metric has a natural connection with Fourier series estimation through Parseval's theorem, which concerns a variant of the weighted quadratic deviation. This simple relationship underlies the computational efficiency of the optimizing algorithms described in the next chapter. The MISE also is a good choice for comparing different procedures because it is widely accepted by curve estimation researchers who have conducted extensive studies based on this criterion. And, as shown in Figure 3.1, the MISE provides a global assessment of the error by integrating over the entire length of the curve. Thus, the MISE, especially in its unweighted form, is a good choice for getting an overall view of a curve.

However, despite its primacy in the mathematical literature, the unweighted MISE cannot be said to serve every statistical purpose. For example, one particular subregion of the curve may be more important than others. A researcher studying long term survival characteristics may be willing to trade some accuracy in the estimation of the left side of the survival function to improve the accuracy in the right tail. The weighted version of the MISE allows the researcher to do just that through an appropriate choice of the weight function w.

3.2 Fourier series density estimation

This section will deal with the basic problem of estimating a probability density function, f, from a sample of univariate, iid data, $\{X_j\}$. We will

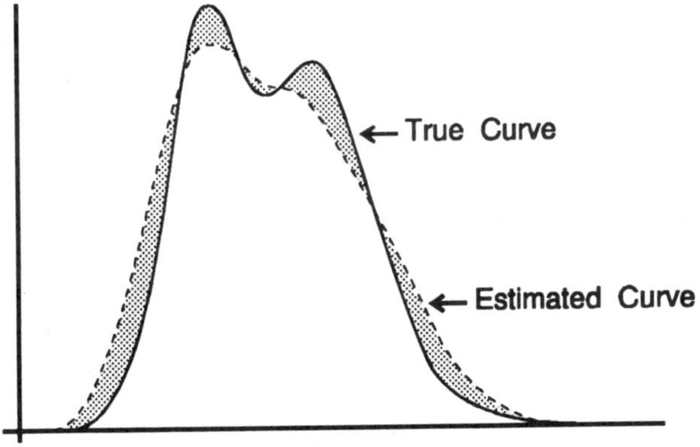

Figure 3.1. *Graphical depiction of the MISE. The shaded area represents the error between the true and estimated curves. The error is squared and integrated over the support of the curves to obtain the MISE.*

assume f is an L_2 function with finite support on [0,1]. Although, technically, f is sometimes treated as a function that is periodic, we will avoid the necessity of defining f in this way by assuming that only values assumed over [0,1] are of interest. A simple scaling transformation will be described in Section 4.8 for densities with support on some other finite interval. In this setting, the Fourier series estimator of the density f can be defined

$$\hat{f}(x)=\sum_k b_k\hat{B}_k\exp\{2\pi ikx\}, \tag{3.2.1}$$

where $\hat{B}_k$, $k=0,\pm1,\pm2\ldots$, the sample Fourier coefficients, are defined

$$\hat{B}_k = n^{-1}\sum_{j=1}^{n}\exp\left\{-2\pi ikX_j\right\},\quad k = 0,\pm1,\pm2,\ldots. \tag{3.2.2}$$

Here b_k, $k=0,\pm1,\pm2,\ldots$ can either be referred to as equally spaced KT evaluations, or by the more compact designation, multipliers.

There is an advantage to conceptualizing multipliers in terms of some KT curve. A curve is usually thought of as a single entity, while the word multipliers is a plural form. As a plural it can lead to a researcher's losing sight of the interconnectedness of members of some multiplier sequence. Since they are aligned with nut and bolt details of curve estimation computational processes in this and the next chapter, we will use the term multipliers. However, because it facilitates the geometric depiction of multiplier behavior and properties, in Chapter 5 we will use both the multiplier and KT formulations of this basic tool.

Figure 3.2 is a listing of a FORTRAN function that calculates an estimate of the kth sample Fourier coefficient. In practice, this routine would be called several different times with consecutive values of k in order to generate enough coefficients to estimate the density adequately. The question of exactly how many coefficients are needed is addressed in the next chapter. Note that the FORTRAN routine does not calculate $\hat{B}_k$ exactly as defined in (3.2.2) but instead relies on Euler's well known formula for the trigonometric form of a complex exponential:

$$e^{iz} = \cos z + i\sin z. \tag{3.2.3}$$

```
      complex function kCoef (k, X, n)
*********************************************
*
* Estimates the k-th Fourier coefficient from
* the points X(1),...,X(n)

* Declarations:
      integer    k, n
      real       X(n)

* Local Declarations:
      integer    j
      real       rpart, ipart

* Constant Declarations:
      real       PI
      parameter (PI=3.1415927)

      rpart = 0.0
      ipart = 0.0

      do 100, j=1, n
        rpart = rpart + cos( 2*PI*k * X(j) )
        ipart = ipart + sin( 2*PI*k * X(j) )
100     continue

      kCoef = cmplx( rpart, -ipart ) / n

      return
      end
```

Figure 3.2. *FORTRAN function for estimating the kth Fourier coefficient.*

This choice was made for reasons of efficiency. Calls to cosine and sine functions are usually executed much more quickly than the computational exponentiation of a complex number. Euler's identity makes it easy to see that $\hat{B}_k^*$, the complex conjugate of $\hat{B}_k$, is equal to $\hat{B}_{-k}$. This fact can also be used to streamline density estimation algorithms. Only coefficients with positive indices need to be estimated from the data, since $\{\hat{B}_k\}$, where k is negative, can then be obtained through simple conjugation.

The sample Fourier coefficients are unbiased estimators of f's Fourier coefficients. The proof that $\hat{B}_k$ is an unbiased estimate of B_k is straightforward:

$$\mathrm{E}\left[\hat{B}_k\right] = n^{-1}\sum_{j=1}^{n} \mathrm{E}\left[\exp\left\{-2\pi ikX_j\right\}\right]$$
$$= n^{-1}\sum_{j=1}^{n} \int_0^1 f(x)\exp\{-2\pi ikx\}dx = \hat{B}_k.$$

Unlike the unbiased estimates of its Fourier coefficients, the unbiased estimation of f itself is not straightforward. If the Fourier expansion of f includes an infinite number of terms, then to construct an unbiased estimator, an infinite number of coefficients must be estimated. And even if f only requires a finite expansion, the exact number of terms that must be estimated is still not known with certainty. Thus, it is likely that $\hat{f}$ will be a biased estimator of f. In fact, Rosenblatt (1956) has shown that all generalized density estimates satisfying certain mild conditions are biased.

In mathematical terms, the sample Fourier coefficients result from the mapping of data into the Fourier domain, or as it is more commonly called, the frequency domain. For many applications the sample coefficients encapsulate virtually all vital information contained in the original sample. This means that once the coefficients have been calculated for the density estimate, there is no need to return to the data to estimate other statistics or curves such as the survival or $R(x)$ functions.

One reason for the computational advantage of the Fourier-based approach is that the number of series terms, m, required to estimate a curve is commonly far smaller than the number of original data points. For example, the density function estimated from a sample of $n=1000$ normally distributed points can usually be adequately represented with m sample Fourier coefficients, where m is much smaller than 1000 (usually $m<7$). Although the reduction in data is less pronounced for complex distributions such as those discussed in Chapters 5 and 6, even in these cases m is usually much smaller than n. For example, in the above normal example the standard deviation of the distribution can be estimated with approximately 20 (complex) operations in the Fourier domain as opposed to 1000 operations with the original data.

By performing a large variety of roles within a common framework, multipliers provide a unified approach by which many exploratory curve estimation procedures can be implemented. For example, the limits of the summation were deliberately left unspecified in the density estimator (3.2.1) defined above. The setting of summation limits is the equivalent of determining which consecutive multipliers take on the value zero as the absolute values of their indices increase. However, multipliers can perform far more sophisticated statistical and mathematical purposes than simply replacing summation limits. Their primary role is the determination of which curve to estimate. That is, in an interactive data analysis session, the same set of sample Fourier coefficients can be used repeatedly, while many different curves or composites of curves can be constructed by choosing different multipliers.

That one wishes to estimate a pdf f and not a cdf F is one form of curve estimation decision. That one wishes to set the weight $w \equiv 1$ and not choose w to be an increasing function of x is another decision. Choosing a contrast enhanced curve rather than a smoothed curve is a third decision. The primary advantage of multiplier-based implementation is that these three, together with many other curve estimation processes, can be placed within a single analytical and computational framework.

Substituting the multiplier sequence

$$b_k = 1, \ |k| \leq m; \ b_k = 0, \ |k| > m \tag{3.2.4}$$

into the estimator defined by (3.2.1) results in what Wahba (1981) calls the "raw" Fourier series density estimator:

$$\hat{f}_R(x) = \sum_{k=-\mathrm{m}}^{m} \hat{B}_k \exp\{2\pi ikx\}. \tag{3.2.5}$$

Thus,the default or neutral choice of $\{b_k\}$ is the choice that corresponds to an estimate of the probability density function.

Figure 3.3 is a listing of a FORTRAN function called "pdf", which evaluates the raw density estimator at a point x. The routine assumes that at least m coefficients have been estimated elsewhere and are stored in the complex array B. Like the coefficient-generating routine described earlier, pdf employs the trigonometric form of complex exponentials

```
      real function pdf (x, B, m)
**********************************************
*
* Evaluates the raw Fourier series estimator
* at the point x.  Assume coefficents have
* been estimated and are stored in the
* complex array B(0).....B(m)

* Declarations:
      integer    m
      real       x
      complex    B(0:m)

* Local Declarations:
      integer    k
      real       temp

* Constant Declarations:
      real       PI
      parameter (PI=3.1415927)

      temp = 0.0

      do 100, k=1, m
        temp =  temp
     $       +  real(B(k)) * cos( 2*PI*k*x )
     $       - aimag(B(k)) * sin( 2*PI*k*x )
100     continue

      pdf = 1 + 2*temp

      return
      end
```

Figure 3.3. *FORTRAN function for calculating density estimate.*

and also exploits the identity $\hat{B}_{-k} = \hat{B}_k^*$ to increase the computational efficiency of the algorithm. The routine also treats the real and imaginary parts of $\hat{B}_k$ individually to eliminate redundant operations. If $\hat{B}_k$ is written $\hat{C}_k + i\hat{S}_k$, then

$$\hat{f}_R(x) = \hat{B}_0 + \sum_{k=1}^{m} \left\{ \hat{B}_k e^{2\pi ikx} + \hat{B}_{-k} e^{-2\pi ikx} \right\}$$

$$= 1 + \sum_{k=1}^{m} \left\{ \hat{C}_k \left(e^{2\pi ikx} + e^{-2\pi ikx} \right) + i\hat{S}_k \left(e^{2\pi ikx} - e^{-2\pi ikx} \right) \right\}$$

$$= 1 + 2\sum_{k=1}^{m} \left\{ \hat{C}_k \cos 2\pi kx - \hat{S}_k \sin 2\pi kx \right\}.$$

The curves shown in Figure 3.4 are raw estimates of a Weibull density with shape parameter 2.2 scaled and located within the [0,1] interval by transforming maximum and minimum sample values to 1/3 and 2/3 for each sample. Each curve was created using a different value of m and each required 100 evaluations of the pdf routine.

Different choices for $\{b_k\}$ lead to estimates of different curves. One example is the first derivative of the density, f'. Although not as familiar to statisticians as the density or cumulative functions, the estimated derivative, $\hat{f}'$, does have several applications in the area of curve estimation. For example, the zeroes of $\hat{f}'$ provide estimates of the local density function modes. Applications of the estimated density second derivative have been described by Silverman (1978) and Scott, Tapia and Thompson (1979).

To estimate f' in many applications the raw density estimate can be differentiated term by term. One advantage of uniformly convergent Fourier series representation is that differentiation and integration are simple term by term operations:

$$\hat{f}'(x) = \frac{d}{dx}\hat{f}(x) = \sum_{k=-m}^{m} \frac{d}{dx}\hat{B}_k \exp\{2\pi ikx\}$$

$$= \sum_{\substack{k=-m \\ k\neq 0}}^{m} (2\pi ik)\hat{B}_k \exp\{2\pi ikx\}.$$

Therefore the first derivative of the density can be estimated by substituting the sequence

$$b_k = 2\pi ik, \ \ |k| \leq m, k\neq 0; \ \ b_k = 0, \ \ |k| > m, \ k=0, \qquad (3.2.6)$$

into (3.2.1). This example shows how a particular multiplier sequence can be applied to move from estimating one curve to estimating another.

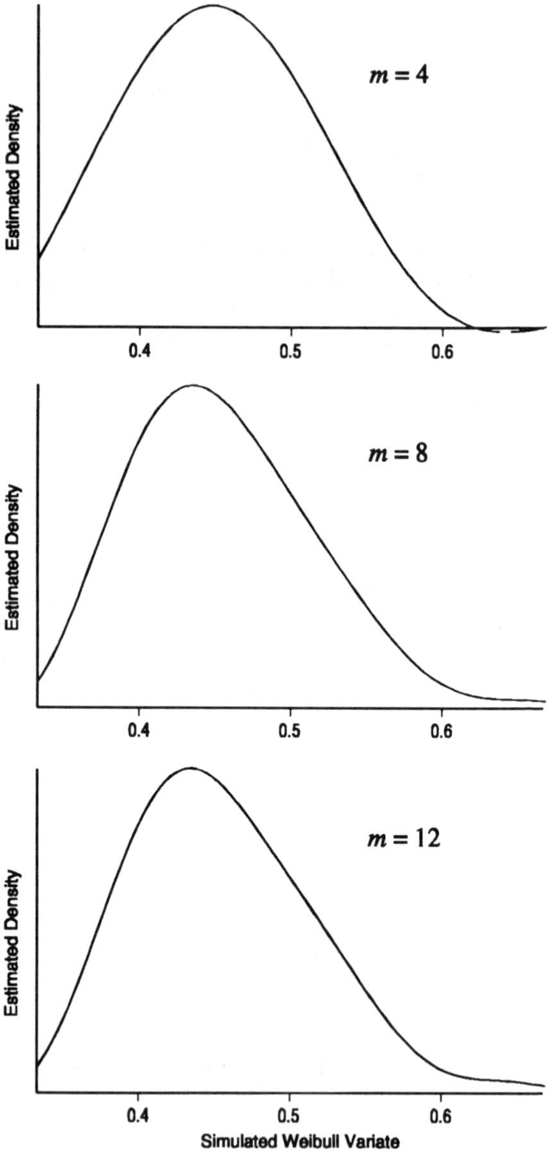

Figure 3.4. *Three raw estimates of a Weilbull density with shape parameter 2.2. Each curve was created using a different value for m, the number of sample Fourier coefficients included in the raw estimator.*

It also illustrates that the field of density estimation can often be extended into the wider realm of curve estimation. A second example shows that the multiplier sequence can be used to estimate certain properties as well as other curves. Define $\hat{\mu}$ as

$$\hat{\mu} = \int_0^1 x\hat{f}(x)dx.$$

Since $\hat{f}(x)$ is a finite sum, it can be integrated term by term:

$$\hat{\mu} = \hat{B}_0 \int_0^1 x dx + \sum_{\substack{k=-m \\ k \neq 0}}^{m} \hat{B}_k \int_0^1 x \exp\{2\pi ikx\} dx. \tag{3.2.7}$$

When $k \neq 0$ the last integral on the right of (3.2.7) can be integrated by parts to yield

$$\begin{aligned}\int_0^1 xe^{2\pi ikx} dx &= \left.\frac{xe^{2\pi ikx}}{2\pi ik}\right|_{x=0}^{1} + \frac{1}{2\pi ik}\int_0^1 e^{2\pi ikx} dx \\ &= \frac{e^{2\pi ik}}{2\pi ik} = \frac{1}{2\pi ik}.\end{aligned} \tag{3.2.8}$$

Noting that $\hat{B}_0 \equiv 1$, (3.2.7) and (3.2.8) can be combined to yield

$$\hat{\mu} = 1/2 + \sum_{\substack{k=-m \\ k \neq 0}}^{m} \frac{1}{2\pi ik} \hat{B}_k.$$

Thus, $\hat{\mu}$ can easily be calculated by first substituting the multiplier sequence $b_k = (2\pi ik)^{-1}$ into estimator (3.2.1), then evaluating the resulting estimator at $x = 0$, and finally adding $1/2$ to the total. Although this procedure is not useful in itself, we will see in Chapter 8 that it can be generalized to provide a valuable form of nonparametric regression.

While the estimates of the density, its derivative, and $\hat{\mu}$ only require one summation, a third example involves two summations, one of which has $\hat{\mu}$ as a limiting case. Like procedures for estimating the density and its derivative, the estimation of the cdf, F, can be based on the choice of a specific multiplier sequence applied to the same set of sample

Fourier coefficients that were used in previously described applications. Following a strategy similar to that used to derive $\hat{f}'$, here we integrate the raw estimator, $\hat{f}(x)$, term by term to estimate F:

$$\begin{aligned}\hat{F}(x) &= \hat{B}_0 x + \sum_{\substack{k=-m\\k\neq 0}}^{m} \hat{B}_k \int_0^x e^{2\pi i k y}\,dy \\ &= x + \sum_{\substack{k=-m\\k\neq 0}}^{m} \frac{\hat{B}_k}{2\pi i k}\left[e^{2\pi i k x} - 1\right] \\ &= x - \sum_{\substack{k=-m\\k\neq 0}}^{m} \frac{\hat{B}_k}{2\pi i k} + \sum_{\substack{k=-m\\k\neq 0}}^{m} \frac{\hat{B}_k}{2\pi i k} e^{2\pi i k x} \\ &= x - \hat{\mu} + \frac{1}{2} + \sum_{\substack{k=-m\\k\neq 0}}^{m} \frac{\hat{B}_k}{2\pi i k} e^{2\pi i k x}.\end{aligned}$$

Again we find that multiplier sequence defined by $b_k = (2\pi i k)^{-1}$ plays the role of delineating the targeted curve.

Up to this point we have employed multipliers for only mathematical or computational purposes. However, $\{b_k\}$ also plays a role that is primarily statistical in nature. Multipliers can be chosen on the basis of data usage efficiency. Many sections of this book concern the design or modification of a multiplier sequence to ensure that it minimizes either the weighted or unweighted MISE. Traditionally, methods for optimizing Fourier series density estimators have employed one of two basic strategies. The first simply entails the careful choice of the point m at which the raw estimator defined by (3.2.5) is truncated. Criteria for truncation point determination are usually called "stopping rules." Truncating the raw estimator is equivalent to choosing one particular multiplier sequence from among alternative sequences consisting exclusively of consecutive values 0 or 1, as was done in sequence (3.2.4). This sequence was first introduced as only a computational and representational convenience. However, despite its extreme simplicity, since the choice of truncation point is critical to the production of any accurate series-based curve estimate, we will consider it below in detail.

Unlike the mathematical process of approximating a function through a truncated expansion, the inclusion of an increasing number of terms in (3.2.5) does not necessarily result in a better density estimate. For example, Crain (1973) showed that the choice of multiplier sequence $b_k = 1$, $k = 0, \pm 1, \pm 2, \ldots$, that is, an untruncated raw estimator, leads to estimates with infinite MISE. This implies that the inclusion of some finite number of terms m will be optimal with respect to minimizing the MISE of $\hat{f}$.

A stopping rule depends on essentially one parameter—the number of terms included in the estimator—to control how well $\hat{f}$ estimates the true density. In contrast, the second optimization strategy allows b_k to assume any real value. By choosing multipliers that eventually decrease as k increases, the coefficients of $\hat{f}$ can be gradually tapered to zero instead of being sharply truncated. This strategy offers potentially far greater control over the optimization process than does the pure raw estimator truncation strategy.

3.3 Kernel density estimation

There are several advantages of the Fourier series representation of curve estimators, one of which is the computational modularity illustrated in the previous section. Another advantage is the ease with which Fourier series and transform methods can be extended to a multivariate setting.

A third benefit is the near identity between Fourier estimators based on $\{\exp(2\pi ikx)\}$ and another extensively studied class of procedures—kernel estimators. With few exceptions a particular estimator may be viewed as resulting from either a Fourier series or a kernel estimation process. Just as the association of multipliers with the KT curve allows a great range of applications and possibilities, the duality of series and kernel representation allows a researcher to choose whichever interpretation is more useful for a specific application. This close correspondence means that theoretical results derived for one form of representation are almost always applicable to the other method as well. The kernel form of the estimator is often a helpful tool for the visualization of certain series estimation processes that are computationally convenient, but difficult to interpret when viewed entirely from a series perspective.

The best way to introduce kernel representation is first to take a look at a closely related form of curve. Although widely used as an exploratory tool in data analysis, the histogram is not always recognized as a type of density estimator. But with only one significant alteration the histogram would indeed be a crude kernel density estimator.

The construction of the histogram is depicted graphically in Figure 3.5. As a first step the interval of the real line encompassing the data is divided into equal-sized segments called bins. In the figure bins are marked by dashed lines. The next step is to count the number of points, f_1, that fall within the first bin. Then f_1 blocks are stacked in the bin. The procedure is then repeated for each of the other bins. Note that since a two-dimensional block is necessarily wider than a zero dimensional point, the histogram performs the role of spreading the influence of any one point along and over a small portion of the variate axis. Although spreading is a common feature of many density estimation procedures, the manner in which it is done can differ greatly between methods.

The kernel estimator of the density generalizes the basic histogram in two significant ways. The first addresses the arbitrary manner in which the influence of a point is spread within its bin. Note that some points in Figure 3.5 are located near the left boundary of their bins while others are closer to the right. On points near the left side of a bin, the majority of the block placed above them is to their right; the converse is true for the points close to the left side. Thus, owing to the arbitrary placement of the bins, the influence of most of the points is spread asymmetrically along the real line.

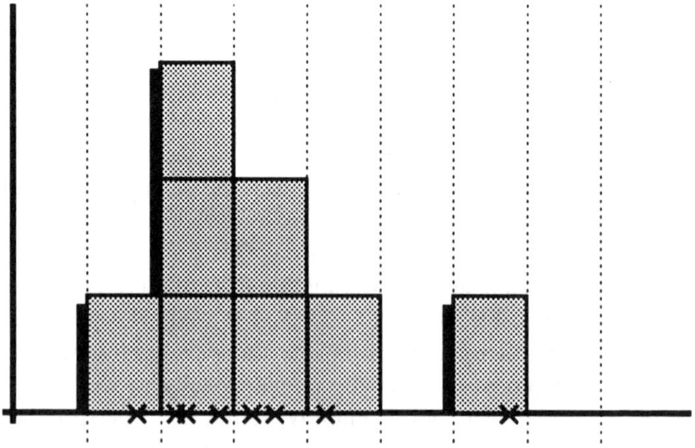

Figure 3.5. *Construction of the histogram. Each X represents the location of a data point. The dotted lines delineate the boundaries of the individual bins.*

When a histogram is drawn by hand, or when it is actually built from real blocks, the division of the data space into bins is a necessary step. But, when a computer algorithm is used, binning or the allocation of points to class intervals is unnecessary. The chief difference between a histogram and a kernel estimate is that the kernel process places the center of the block directly over the data point. Thus the influence of a point is spread equally around it. An example of this type of estimator, called the "naive" estimator, which differs from the histogram only by the way the block is centered, can be found in Rosenblatt (1956).

The second modification of the simple histogram construction process underlies the generality of kernel-based estimation. What were once blocks become kernels when they are allowed to assume a greater variety of shapes than the rectangles shown in Figure 3.5. Thus, the word block would be a misnomer when used in reference to a kernel estimator unless some kernel happens to be a rectangle. Several popular choices for kernel functions are shown in Figure 3.6. Note that one is rectangular, the choice that corresponds to the naive estimator described above. The construction of a kernel estimate using gaussian kernels is depicted in Figure 3.7. A small gaussian-shaped curve is centered directly over each data point and then the contribution from each individual kernel is summed to create the overall estimate of the density.

The added flexibility of the kernel estimation process also brings with it more decisions for the researcher, one of which is: which kernel $K(z)$ to use? Surprisingly, many density estimation researchers do not consider kernel shape to be of crucial importance, because, from the point of view of minimizing the MISE, they feel that there is not much difference between commonly chosen kernels. For example, with respect to MISE, the worst function shown in Figure 3.6 is the rectangular kernel, which is only about 7% less efficient than the best, the Epanechnikov kernel (Silverman, 1986). For this reason, kernel shape is commonly chosen solely on the basis of ease of implementation or computational efficiency criteria.

After a kernel's shape is selected, how wide each kernel should be still must be decided; in other words: how widely should the influence of each data point be spread? Intuitively it seems preferable for a kernel to be narrow when the true density is narrow or sharply peaked, and wide when the true density is flatter. In the kernel literature the process of making this determination is commonly referred to as "smoothing." However, for reasons that will be detailed in Chapter 5, frequently we will substitute the word optimization for the word smoothing. Optimiza-

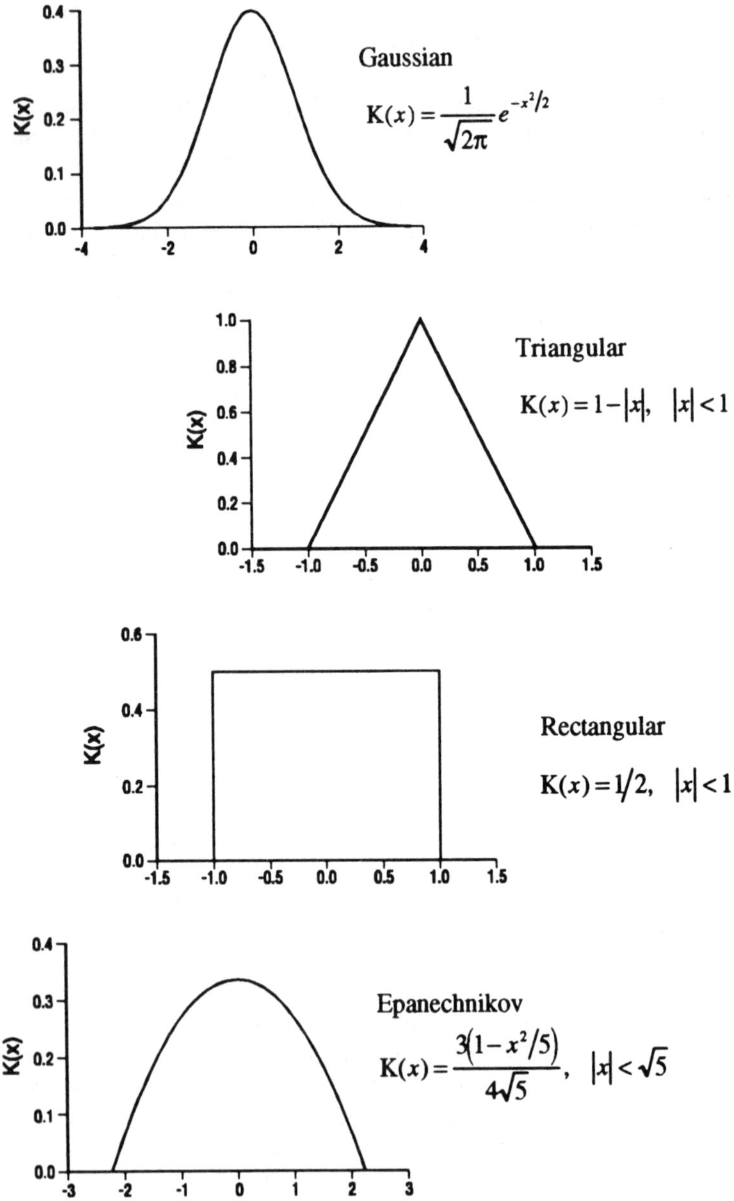

Figure 3.6. *Selected kernel functions.*

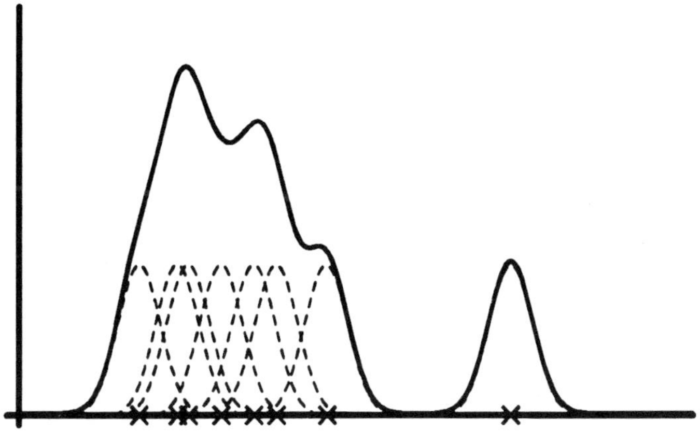

Figure 3.7. *Construction of the kernel estimator. Gaussian-shaped kernels (dashed curves) are centered over the same data points shown in Figure 3.5. The kernel estimate (solid curve) is then made by summing the contributions of each individual kernel.*

tion is a preferable designation because the smoothness of a curve may have little bearing on its accuracy and, in many cases, smoothness itself is a form of distortion. Sometimes the best or optimal estimate is much rougher than its smoothed variant.

Note that smoothing is also an important, though seldom emphasized, aspect of histogram construction. Although it was not mentioned in the description of Figure 3.5, the degree of smoothness of this histogram was determined implicitly by the allocation of the sample among the bins commonly called class intervals. Choice of a large bin width will produce a smoother, flatter, histogram, while a narrow bin choice will have the opposite effect. In this respect the bin width of a histogram is analogous to the spread or width of the kernel.

Mathematically, the kernel estimator can be defined as

$$\hat{f}(x) = \frac{1}{nh}\sum_{j=1}^{n} \mathrm{K}\left(\frac{x-X_j}{h}\right), \tag{3.3.1}$$

where K is the kernel function and h is called the bandwidth or smoothing parameter. K is commonly selected a priori from a class of nonnegative and symmetric elementary functions such that

$$\int K(x)dx=1 \text{ and } \int xK(x)dx=0.$$

A function constructed by using a kernel that satisfies these conditions will be a density. In many instances K has been chosen to be a function such as the gaussian density, which is often encountered in parametric forms of statistical analysis, as shown in Figure 3.6A:

$$K(x) = \frac{1}{\sqrt{2\pi}\, h} \exp\{-x^2/2h^2\}.$$

Note that the symbol for the scale parameter of the gaussian density, σ, has been replaced here by the smoothing setting symbol, h. Thanks to the 2 after the division slash within $\{-x^2/2h^2\}$, the gaussian scale parameter equals the property, standard deviation. As the smoothing parameter h increases, or equivalently, as the standard deviation of the gaussian kernel increases, the influence of each datum is spread farther and farther from the point.

Another common assumption made when kernel estimates are applied is that the bandwidth is a function of the sample size n satisfying

$$h_n \to 0 \text{ and } nh_n \to \infty \text{ as } n \to \infty.$$

These conditions imply that the bandwidth must decrease as n increases, but it should not do so too rapidly. Under these and some other very mild conditions, Parzen (1962) showed that the kernel estimate converges to the true density pointwise in probability as $n \to \infty$. Many more asymptotic results concerning kernel estimates may be found in the book by Prakasa Rao (1983).

Although not immediately obvious, a kernel curve estimator can be expressed over any finite interval as a Fourier series estimator. This can be seen through a simple rearrangement of the terms in expression (3.2.1):

$$\hat{f}(x) = \sum_k b_k \hat{B}_k \exp\{2\pi ikx\}$$

$$= \sum_k b_k \left[n^{-1} \sum_{j=1}^{n} \exp\left\{ -2\pi i k X_j \right\} \right] \exp\{2\pi i k x\}$$

$$= n^{-1} \sum_{j=1}^{n} \left[\sum_k b_k \exp\left\{ 2\pi i k (x - X_j) \right\} \right].$$

If $b_k = b_{-k}$ and if $\{b_k\}$ forms a convergent series, then $\{b_k\}$ can be considered as the Fourier coefficients of the kernel function defined by the Fourier series density estimator, that is, for $h=1$,

$$\mathrm{K}(x) = \sum_k b_k \exp\{2\pi i k x\}. \tag{3.3.2}$$

Alternatively, by using the expressions for the characteristic function of truncated densities given by Kronmal and Tarter (1968), many kernel estimators can be expressed as Fourier series estimators. These facts underlie the useful conceptualization that multipliers are equivalent to the equally spaced KT evaluations described in the previous chapter.

As an illustration of the duality of Fourier series and kernel methodology, consider the raw estimator defined in Section 3.2. The multiplier sequence for the raw estimator consisted of only 1's and 0's; specifically $b_k = 1$ for all k between $-m$ and m, and $b_k = 0$ elsewhere. The kernel defined by the raw estimator can be derived by substituting this multiplier sequence into (3.3.2):

$$\begin{aligned} \mathrm{K_D}(x) &= \sum_{k=-m}^{m} 1 \cdot e^{2\pi i k x} = \sum_{k=0}^{2m} e^{2\pi i (k-m) x} \\ &= e^{-2\pi i m x} \sum_{k=0}^{2m} e^{2\pi i k x} \\ &= e^{-2\pi i m x} \left[\frac{e^{2\pi i (2m+1) x} - 1}{e^{2\pi i x} - 1} \right], \end{aligned} \tag{3.3.3}$$

where the final step above is obtained through the summation formula for finite geometric series. A final simplification to (3.3.3) results from the following identity which relies on Euler's formula for complex exponentials:

$$\begin{aligned} e^{2\pi iz}-1 &= (\cos\pi z + i\sin\pi z)^2-1 \\ &= 2i\sin\pi z\cos\pi z - 2\sin^2\pi z \\ &= 2i\sin\pi z(\cos\pi z + i\sin\pi z) \\ &= 2i\sin\pi z\, e^{\pi iz}. \end{aligned} \tag{3.3.4}$$

Two applications of (3.3.4), one with $z = (2m+1)x$ and the second with $z = x$, yield new expressions for the numerator and denominator of (3.3.3):

$$\begin{aligned} K_D(x) &= e^{-2\pi imx}\left[\frac{2i(\sin\pi(2m+1)x)e^{\pi i(2m+1)x}}{2i(\sin\pi x)e^{\pi ix}}\right] \\ &= \frac{\sin\pi(2m+1)x}{\sin\pi x}. \end{aligned}$$

The expression above defines the famous Dirichlet kernel of Fourier analysis (Kronmal and Tarter, 1968; Dym and McKean, 1972).

The duality of Fourier series and kernel methodology implies that a kernel estimator can be constructed that is mathematically equivalent to the raw Fourier series estimator by selecting the Dirichlet kernel a priori and employing formula (3.3.1) with $h \equiv 1$. Figure 3.8 illustrates the behavior of the Dirichlet kernel for increasing values of m, that is, the kernel determined by increasing the number of terms included in the raw estimator. Note that as m increases, the bulk of the kernel's mass becomes more and more focused on an increasingly narrow region about the data point. A similar type of focusing can be accomplished when using more traditional kernels, such as the gaussian or triangular kernels, by decreasing the smoothing parameter h. Recalling that the smoothing parameter must decrease as sample size increases for the kernel estimator to be a consistent estimator over a point x, it is not surprising that asymptotic results for the raw series estimator usually require that the number of terms tend to infinity as the sample size increases (Hart, 1985).

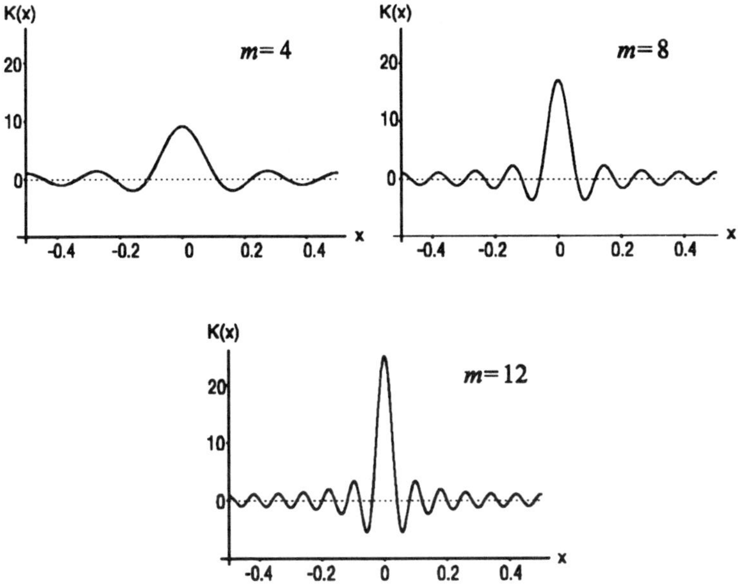

Figure 3.8. *Graph of the Dirichlet kernel for three different values of the parameter m.*

As this example demonstrates, there is nearly perfect theoretical identity between kernel and Fourier series estimators. Still, practical differences between the two methodologies do exist. Some of these differences are due to the peculiarities intrinsic to either kernel or series representation. For example, some kernels have support over the entire real line, while, as noted in Section 3.2, Fourier series estimates are restricted to finite intervals.

Other differences, however, are more perceived than real and often stem simply from notational inconsistencies or misconceptions about the capabilities of each method. One example is the frequently cited misconception that kernel estimates are always nonnegative while series estimates can take on negative values. Yet kernel estimates are guaranteed to be nonnegative only if the kernel function itself is restricted to be nonnegative. Kernel estimates with this restriction relaxed, and which can take on negative values, have been investigated by Müller (1984).

On the other hand, nonnegative estimates have rarely been emphasized by series researchers. One could, however, ensure that a Fourier series estimate be nonnegative by requiring the multiplier sequence to satisfy

$$\sum_{k,j} b_{k-j} c_k c_j^* \geq 0,$$

for any square-summable sequence $\{c_k\}$ (Anderson and de Figueiredo, 1980). In particular, the multiplier sequence

$$b_k = 1 - \frac{k}{m+1}, \quad |k| \leq m,$$
$$b_k = 0, \quad |k| > m,$$

where m is a positive integer, was suggested by Kronmal and Tarter (1968). The use of this sequence is equivalent to the selection of a Fejér kernel, which can be expressed

$$K_F(x) = \frac{1}{m+1}\left[\frac{\sin\pi(m+1)x}{\sin\pi x}\right]^2$$

(Dym and McKean, 1972). Figure 3.9 shows, that, although the Fejér kernel oscillates like the Dirichlet kernel, the Fejér kernel is everywhere

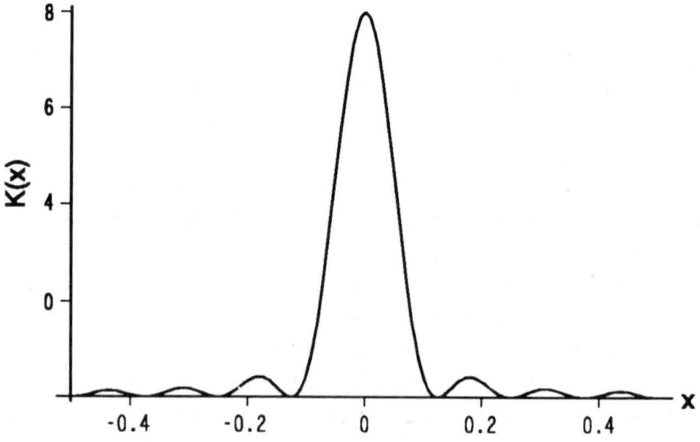

Figure 3.9. *Graph of the Fejér kernel with the parameter m=8.*

nonnegative. Therefore an estimator constructed with the Fejér kernel will also be nonnegative. Hence, for both series and kernel methodology, estimator nonnegativity is within the control of the investigator and is not a condition intrinsic to either representation.

There is, however, a subtle distinction between the kernel and Fourier series approaches. In the kernel estimation process the smoothing parameter h is explicitly defined to be a multiplicative scaling factor of the kernel function K. Thus, K and h have distinct roles: K controls the shape of the kernel, while h determines its spread. This segmentation of responsibility has led to a widely practiced two-step estimation process: first, a kernel function is selected from some family of curves, for example, a gaussian or triangular density, and second, a single smoothing parameter is chosen based on the characteristics of the data set. Examples of this two-step process may be found in Rudemo (1982), Bowman (1984) and Scott and Factor (1981).

Likewise, the smoothness of the raw Fourier series estimator is also controlled by one parameter—the number of terms included in the estimator. In that situation, however, the parameter determines the spread of the kernel in a more complicated manner than by a simple rescaling. As demonstrated by Figure 3.8, a change in the number of terms, m, not only changes the spread of the kernel but also changes its shape slightly. Note that as m increases, the amplitude in the oscillations in the function increase as well. And, unlike simple kernel estimators, the goodness-of-fit of the series estimators need not be determined by only a single parameter but can instead be governed by the entire sequence of multipliers.

This lack of dependence on a single explicitly specified smoothing parameter has allowed researchers to experiment with a wide variety of strategies for controlling the smoothness of an estimate. For example, Tarter and Kronmal (1976), Diggle and Hall (1986) and Hart (1985) have all suggested data-dependent methods for choosing an optimal number of terms to include in the raw Fourier series estimator. The detailed implementation of these methods will be described in Chapter 4. Alternatively, Wahba (1981) has suggested a multiplier sequence dependent on two parameters:

$$b_k = \frac{1}{(1 + \lambda(2\pi k)^{2p})}, \quad \lambda \geq 0, p > 1/2, |k| \leq n/2;$$
$$b_k = 0, \quad |k| > n/2.$$

The Wahba procedure first fixes the number of terms included in the density estimator and then minimizes an estimator of the MISE to find suitable values for the parameters and p.

Another example is the sequence proposed by Brunk (1978):

$$b_k = \frac{n}{(n-1+\pi_k)},$$

where π_k is a measure of precision based on prior information. Brunk notes that the kernel resulting from the use of this sequence depends in a complicated manner on the specification of a prior distribution.

Finally, rather than controlling smoothness through one, two, or any fixed number of parameters, Watson (1969) and Fellner and Tarter (1971) have considered estimating each term of the multiplier sequence individually. Since the multiplier sequence also defines a particular kernel, these methods estimate the shape, spread, and all other aspects of the kernel simultaneously.

CHAPTER 4

Optimizing density estimates

Multipliers can be applied to the important process of optimizing a density estimator with respect to some metric or measure of closeness. Loosely speaking, the multiplier sequence can be used to fine-tune the raw Fourier series estimator in the sense of obtaining estimates that are close to the true density in some well-defined sense. In this chapter we will discuss several different methods for choosing optimizing multipliers with regard to the weighted mean integrated square error (MISE) metric:

$$J\left(\hat{f},f,w\right) = \mathrm{E}\int_0^1 \left\{f(x)-\hat{f}(x)\right\}^2 w(x)dx. \tag{4.0.1}$$

Here $w(x)$ is a nonnegative weight function, which can be chosen preferentially to emphasize estimation over one subregion of the density relative to estimation over another subregion. For developments concerning other measures of closeness, especially those based on the integrated absolute difference between the estimate and the curve, see Devroye (1987) and Devroye and Gyorfi (1985).

The first several sections of the chapter will consider methods based on the special case of the unweighted MISE where $w(x) \equiv 1$, which we will call below simply the MISE. Among these methods are several stopping rules where each multiplier, b_k, assumes either the value zero or one. A second class of more general methods allows the multiplier sequence to take on a continuous range of real values. Two of these more general sequences will be introduced and their relationship to stopping rules will be discussed. The last section of the chapter concerns the weighted version of the MISE. In particular, we will consider both a stopping rule and a general multiplier sequence that is designed to correspond to a useful class of weight functions.

4.1 Preliminary considerations

Over the years, a myriad of procedures, some more clearly stated than others, have been proposed to optimize density estimators. Those based on a kernel formulation, for example, the method due to Scott, Tapia and Thompson (1977), commonly utilize the formula for the optimal kernel window width published by Parzen (1962). Other methods are directly applicable to both series and kernel estimation and include those based on a combination of likelihood and cross-validation approaches (Duin, 1976; Habbema, Hermans and van der Broek, 1974); the combination of least squares methodology with cross-validation (Rudemo, 1982; Bowman, 1984); the generalized cross-validation criteria of Wahba (1977); and the penalized likelihood criteria of Good and Gaskins (1980). Silverman (1986) and Izenman (1991) provide extensive descriptions and discussion of several of these methods while Monte Carlo comparisons between some of the methods can be found in Park and Marron (1990), Bowman (1985), and Scott and Factor (1981). A general overview of cross-validation methodology is given by Stone (1974).

Some methods for optimal density estimation are especially well suited to series-based representation. These procedures exploit the special relationship between the unweighted MISE and the Fourier series expansions of both the true and estimated densities. Specifically, using Parseval's identity, the MISE can be expressed concisely in terms the Fourier coefficients of the density f and its estimator $\hat{f}$:

$$J\left(\hat{f},f\right) = \mathrm{E}\int_0^1 \left\{\hat{f}(x) - f(x)\right\}^2 dx \qquad (4.1.1)$$

$$= 2\mathrm{E}\sum_{k=1}^{\infty} \left| b_k \hat{B}_k - B_k \right|^2 .$$

Only a few straightforward manipulations are required to forge this simple expression into a multipurpose tool for optimizing multiplier sequences and stopping rules. Expanding the square in (4.1.1) yields:

$$J\left(\hat{f},f\right) = 2\mathrm{E}\sum_{k=1}^{\infty} \left\{ b_k^2 \hat{B}_k \hat{B}_k^* - b_k \hat{B}_k^* B_k + B_k B_k^* \right\}, \qquad (4.1.2)$$

where Z^* represents the complex conjugate of Z and $|Z|^2$, called the squared modulus of Z, is equal to ZZ^*. Consider the terms $\mathrm{E}[\hat{B}_k\hat{B}_k^*]$, and $\mathrm{E}[\hat{B}_k\hat{B}_l^*]$ where

$$n^{-2}\mathrm{E}\left[\hat{B}_k\hat{B}_l^*\right] = \mathrm{E}\sum_{s=1}^{n}\exp\{2\pi ikX_s\}\sum_{\substack{t=1\\ s\neq t}}^{n}\exp\{2\pi ilX_t\}$$

$$+\,\mathrm{E}\sum_{s=1}^{n}\exp\{-2\pi i(k-l)X_s\}$$

$$=\sum_{s=1}^{n}\sum_{\substack{t=1\\ s\neq t}}^{n}\int_0^1 f(x)e^{-2\pi ikx}dx\int_0^1 f(y)e^{2\pi ily}dy \tag{4.1.3}$$

$$+\sum_{s=1}^{n}\int_0^1 f(x)e^{-2\pi i(k-l)x}dx$$

$$= n(n-1)B_kB_{-l}+nB_{k-l}$$

and

$$\mathrm{E}\left[\hat{B}_k\hat{B}_k^*\right] = n^{-1}\left(1+(n-1)\left|B_k\right|^2\right). \tag{4.1.4}$$

(4.1.4) implies

$$\mathrm{Var}\left(\hat{B}_k\right) = n^{-1}\left(1-\left|B_k\right|^2\right). \tag{4.1.5}$$

Here $\mathrm{Var}(\hat{B}_k)$ is defined as $\mathrm{E}[\hat{B}_k\hat{B}_k^*]-|B_kB_k^*|$. Note that the value of $|B_k|^2$ is bounded within [0,1]. Consequently, the variance of $\hat{B}_k$ increases as $|B_k|^2$ decreases. It will be shown in later sections that this property of the sample trigonometric moment variance plays a crucial role in the design of optimization techniques.

The substitution of (4.1.5) into (4.1.1) implies

$$\mathcal{U}\left(\hat{f},f\right) = 2\sum_{k=1}^{\infty}\left\{n^{-1}b_k^2(1-\left|B_k\right|^2)+(1-b_k)^2\left|B_k\right|^2\right\}. \tag{4.1.6}$$

Suppose $\{b_k\}$ contains a subsequence of 1's followed by a subsequence of 0's. When interpreted as equally spaced evaluations of the KT curve introduced in Chapter 2, this $\{b_k\}$ corresponds to the truncated rectangular KT, which will be shown below to result from the application of a series stopping rule. The formula for the MISE of curve estimators based on this KT is

$$J\left(\hat{f},f\right) = 2\sum_{k=1}^{m} n^{-1}\left(1-\left|B_k\right|^2\right) + 2\sum_{k=m+1}^{\infty}\left|B_k\right|^2. \qquad (4.1.7)$$

Hence, the raw estimator MISE, which corresponds to a rectangular KT, can be broken down in a manner very similar to the classical breakdown of ordinary mean square error parameter estimator efficiency criteria into variance and bias components. The first term on the right of (4.1.7) is the sum of the variances of each estimated coefficient included in the estimator. The second term is the sum of the squared moduli of every coefficient not included in the estimator, that is, the sum of the squared biases of the unestimated coefficients. Since every summand on the right of (4.1.7) is nonnegative, it is clear that MISE-based optimization of a rectangular KT is a tradeoff between variance and bias. A longer rectangle that includes more estimated coefficients in $\hat{f}$ decreases bias but increases variance.

The modulus of individual coefficients involved plays a pivotal role in MISE-based KT determination. From formula (4.1.5) it can be seen that Fourier coefficients with larger magnitudes can be estimated more accurately than smaller coefficients, that is, with less variance. Thus, including coefficients with large moduli in $\hat{f}$ only minimally increases the variance component of (4.1.7) while it substantially decreases estimator bias. This conforms to common sense, since the coefficients with the largest magnitude also have the greatest affect on the characteristics of the distribution. Loosely speaking, the MISE is minimized by including only estimates of the most important coefficients.

4.2 Single-term stopping rule

One approach to stopping rule construction is based on the minimization of the estimated MISE taken as a function of m, the truncation point or length of a rectangular KT. Although by using substitutions involving $\hat{B}_k$ the first sum on the right side of (4.1.7) can be estimated straightforwardly, estimation of the second sum is a complicated process. Simply replacing each $|B_k|^2$ with its unbiased estimate is clearly unworkable because the result would include an infinite number of terms. To circumvent this difficulty, Tarter, Holcomb and Kronmal (1967) and Kronmal and Tarter (1968) suggested an approach that does not directly estimate the MISE but instead considers the individual contribution of each estimated coefficient to $J(\hat{f},f)$. Specifically, to minimize the MISE, the estimator should incorporate only those terms whose inclusion decreases $J(\hat{f},f)$. Let $\hat{f}_{s-1}$ be the raw estimator that includes $(s-1)$ estimated Fourier coefficients, that is,

$$\hat{f}_{s-1}(x) = \sum_{k=-(s-1)}^{s-1} \hat{B}_k \exp\{2\pi ikx\}.$$

(For simplicity we will discuss only those coefficients with positive indices, since to ensure that the density estimate will be a real-valued function, $\hat{B}_{-k}$ must be included in the estimator whenever $\hat{B}_k$ is included. Thus, when $\hat{B}_0$ is tallied as a term, $\hat{f}_{s-1}$ consists of $2(s-1)+1$ series terms.) Kronmal and Tarter's method concentrates on assessing the change in the MISE due to the addition of the sth estimated coefficient to $\hat{f}_{s-1}$:

$$\begin{aligned}\Delta J_s &= J(\hat{f}_s,f) - J(\hat{f}_{s-1},f)\\ &= 2n^{-1}(1-|B_s|^2) - 2|B_s|^2\\ &= 2n^{-1}(1-(n+1)|B_s|^2).\end{aligned}$$

Using (4.1.4), an unbiased estimator of ΔJ_s can be constructed:

$$\hat{\Delta J}_s = 2n^{-1}\left\{1 - \frac{(n+1)}{(n-1)}\left[n\left|\hat{B}_s\right|^2 - 1\right]\right\}.$$

The MISE is decreased only if ΔJ_s is negative. Accordingly, Kronmal and Tarter proposed a term inclusion rule: include the sth term only if $\hat{\Delta J}_s$ is negative, or equivalently, only if it satisfies the criterion

$$\left|\hat{B}_s\right|^2 > \frac{2}{n+1}. \tag{4.2.1}$$

In the previous section we noted that estimates of only the most important Fourier coefficients should be included in the raw estimator. The above rule helps make this statement more rigorous by supplying an objective criterion for deciding which coefficients can be considered important.

Tarter and Kronmal (1976), however, point out that the application of the inclusion test to an indefinite number of terms can lead to overly rough estimates with consequently large MISE. Figure 4.1 is a raw estimate of a standard normal density where the inclusion rule was applied to 100 estimated Fourier coefficients. Not counting $\hat{B}_0$, which must always be included if a density estimate is to have total mass one, altogether 11 coefficients passed the test. (For simplicity we will discuss only those coefficients with positive indices since the inclusion of $\hat{B}_k$ implies the inclusion of $\hat{B}_{-k}$.) As might be expected, the first terms in the series were the largest, and, consequently, the first seven coefficients passed the inclusion test. Each of the next seven failed, but surprisingly, $\hat{B}_{15}$, $\hat{B}_{16}$ and $\hat{B}_{17}$ were all large enough to merit inclusion. $\hat{B}_{75}$ was the only other term that satisfied the criterion. The inclusion of these higher-order terms is responsible for the overly rough nature of the curve in shown Figure 4.1.

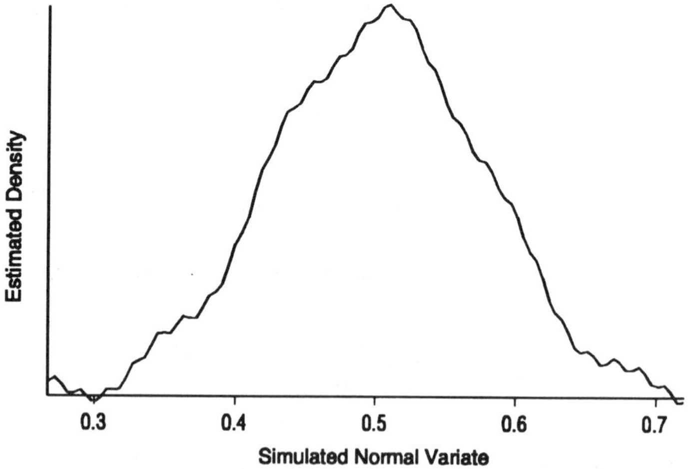

Figure 4.1. *Raw density estimate where the inclusion rule was applied to 100 estimated Fourier coefficients. Data are 1000 standard normal points.*

The simple term inclusion rule is subject to the same problem that underlies any specialized multiple comparisons procedure. Data are reused. Consequently, under the null hypothesis of nonadvisable inclusion, the more terms tested, the more terms will be incorrectly included. As will be discussed in Chapter 10, Watson (1967) showed that if population Fourier coefficients are indeed zero, with increasing sample size the sample trigonometric moments approach independent normally distributed statistics. Hence a half-normal plotting technique can be used for the case of independent normally distributed statistics to "determine which coefficients are numerically too large to have come from zero population values," as described by Hills (1969), based on the work of Daniel (1959).

Despite the simplicity and asymptotic properties of single term stopping rules, such procedures have serious drawbacks. For example, Diggle and Hall (1986) proved that an infinite number of terms are likely to satisfy the single-term stopping rule. Since Crain (1973) showed that an estimator with an infinite number of coefficients has infinite MISE, it is clear that further conditions must be added to the inclusion rule to ensure estimators with good MISE characteristics. Consequently, Tarter and Kronmal (1976) proposed a single-term stopping rule: Include all terms until t consecutive terms have failed to satisfy inequality (4.2.1). (The rule is called single term since the coefficients are tested one at a time.) The authors suggested $t=2$ as a candidate value.

Figure 4.2 is a FORTRAN implementation of the single term stopping rule that begins with $\hat{B}_1$ and continues testing coefficients in ascending order until t consecutive terms fail to satisfy the inclusion criterion. Returning to the above example where a standard normal density is estimated, the single-term stopping rule (with t set to 2) would truncate the estimator after $\hat{B}_7$ since both $\hat{B}_8$ and $\hat{B}_9$ fail the inclusion test. The resulting estimate, shown in Figure 4.3, is clearly superior to the curve shown in Figure 4.1 and illustrates that the inclusion of higher-order terms harmed the previous estimate.

As indicated by the curve in Figure 4.3, the single-term rule can lead to quite accurate estimates of many different types of densities. It is especially accurate when symmetric unimodal densities are estimated. There are some situations, however, in which use of the single-term stopping rule is particularly ill-advised. The problem stems from the choice of consecutive terms t. If, for positive indices, the Fourier coefficients of the true density monotonically decrease, as do the equally spaced evaluations of the normal CF the estimated coefficients tend to

```
      real function stop (B, m, n, t, nterms)
*********************************************
*
* Single term stopping rule.  Returns nterms,
* the number of coefficients which satisfy
* the inclusion criterion. Assume coefficents
* have been estimated and are stored in the
* complex array B(0),...,B(m).  Coefficients
* are tested unitl t terms fail to satisfy
* the criterion.  n is the number or points
* in the original sample.

* Declarations:
      integer  m, n, t, nterms
      complex  B(0:m)

* Local Declarations:
      integer  k, nfail
      real     cutoff

      cutoff = sqrt (2.0/(n+1))
      nterms = 0
      nfail = 0

      do 100, k=1, m
        if (abs(B(k)) .gt. cutoff) then
          nterms = k
          nfail = 0
        else
          nfail = nfail + 1
          if (nfail .ge. t) return
        end if
100   continue
```

Figure 4.2. *FORTRAN function implementing single term stopping rule.*

behave similarly. In such cases, it makes sense to set t to a small number such as one or two. Empirical support for this assertion can be found in the simulation experiment carried out by Diggle and Hall (1986), who tested the single-term rule with different values of t, where $t=1$ or 2 proved the best choice for unimodal normal and beta densities.

On the other hand, if a CF is not monotonic, the single-term rule can be led astray by an incautious choice of t. As an illustration, consider the autoregressive density defined by Hart (1985) as

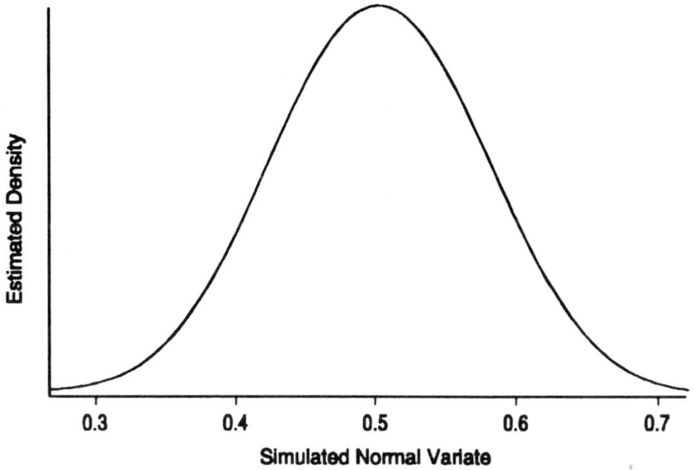

Figure 4.3. *Raw density estimate constructed from the same data described by Figure 4.1, except here, the single-term stopping rule* (*with t*=2) *was used.*

$$f(x) = (0.64)\left|1+(0.6)e^{6\pi ix}\right|^{-2}, \ x \in [0,1], \tag{4.2.2}$$

and graphed in Figure 4.4. As shown in Table 4.1, the Fourier coefficients of this density form a recurrent pattern where each nonzero term

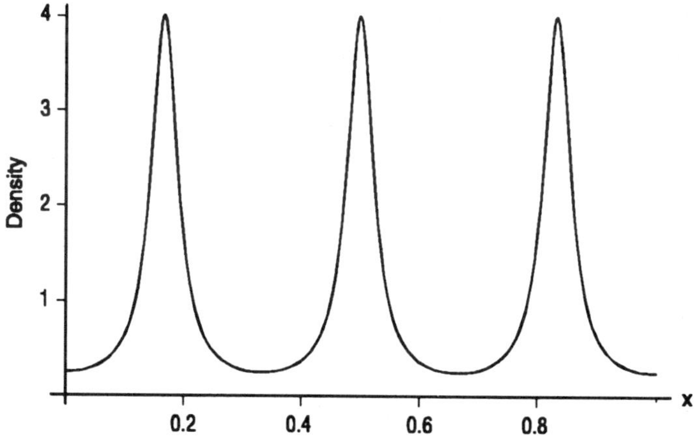

Figure 4.4. *Graph of the autoregressive density function over the interval [0,1].*

Table 4.1. *Fourier coefficients of the autoregressive density.*

k	B_k
0	1
1	0
2	0
3	−0.6
4	0
5	0
6	0.36
7	0
8	0
9	−0.216
10	0
11	0
12	−0.07776

is followed by two consecutive zero terms. If t is set to 2 (or 1), it is possible that the search for included terms will be terminated before later, larger magnitude coefficients are reached. This conjecture is borne out by the results of a simulation conducted by Hart using the autoregressive density. Hart found that in the large majority of trials, the single-term rule, with $t=2$, underestimated the number of terms included in the optimal estimator.

In the case of the autoregressive density, the single-term rule can be improved by setting t equal to a number larger than 2. However, even with this modification the rule could be misleading if three consecutive equally spaced CF zero terms are sandwiched between terms that are larger in magnitude. Obviously, no one choice for t will be right for all situations and, as a consequence, Hart (1985) and Diggle and Hall (1986) have proposed alternative stopping rules which do not rely on t.

4.3 Diggle and Hall stopping rule

Diggle and Hall (1986) have proposed a stopping rule based on combinations of different sample Fourier coefficients. The theory that underlies this stopping rule was developed for general orthogonal series estimators (truncated at m terms) and illustrated through the use of an estimator based on the cosine series:

$$\hat{f}(x) = \sum_{k=0}^{m} \hat{\beta}_k \psi_k(x), \quad x \in [0,\pi],$$

where the sequence of functions

$$\psi_0(x) = \pi^{-1/2},$$
$$\psi_k(x) = (2\pi)^{1/2} \cos kx, \quad k=1,2,\ldots,$$

are orthogonal on the interval $[0,\pi]$, and the sample coefficients are defined as $\hat{\beta}_k = n^{-1}\sum_{j=1}^{n} \psi_k(X_j)$. Below, the essentials of the rule will be detailed specifically as they apply to the Fourier series estimator.

Diggle and Hall's method first constructs an estimator of the MISE, and then conducts a linear search of several candidate values to find the truncation point m that yields the smallest estimated MISE. Presumably, this value of m will correspond to the truncation point that minimizes the true MISE. Thus, the success of this method depends heavily on the MISE estimator's performance.

The estimator of the MISE proposed by Diggle and Hall considers the first and second terms on the right side of (4.1.7) separately. Since the value of any $\hat{B}_k$ is bounded by $[0,1]$, it is ignored and the first term is estimated by $2mn^{-1}$. The second term on the right side of (4.1.7) is more complicated to estimate since it involves the sum of an infinite number of Fourier coefficients. The crucial step of the Diggle and Hall approach is the replacement of an infinite sum by a sum whose limits are allowed to vary. Specifically, these authors note that in many cases, if λ is a function of m such that $\lambda(m) \to \infty$ as $m \to \infty$, then

$$\sum_{k=m+1}^{\lambda(m)m} \left|B_k\right|^2 \sim \sum_{k=m+1}^{\infty} \left|B_k\right|^2$$

as $m \to \infty$. This, together with the formula (4.1.4) for the expected value of $|\hat{B}_k|^2$, suggests the second term of (4.1.7) may be estimated by

$$2(1-n^{-1})^{-1}\sum_{k=m+1}^{\lambda(m)m}\left(\left|\hat{B}_k\right|^2-n^{-1}\right).$$

Combining the estimators of the first and second terms leads to the MISE estimator

$$\hat{J}\left(\hat{f},f\right)=\frac{2m}{n}+\frac{2}{(1-n^{-1})}\sum_{k=m+1}^{\lambda(m)m}\left(\left|\hat{B}_k\right|^2-n^{-1}\right). \tag{4.3.1}$$

Diggle and Hall also offer some practical suggestions regarding the implementation of their stopping rule. For their own simulation study, they chose $\lambda(m)=cm^{1/2}$, where c is a small constant number, and experimented with different values of c. It was found that $c=4$ was a good choice and smaller numbers sometimes produced poor results. With this substitution, (4.3.1) can be expressed

$$\hat{J}\left(\hat{f},f\right)=\frac{2m}{n}-\frac{2m(cm^{1/2}-1)}{(n-1)}+\frac{2}{(1-n^{-1})}\sum_{k=m+1}^{cm^{3/2}}\left|\hat{B}_k\right|^2.$$

Since the quantity $|\hat{B}_k|^2-n^{-1}$ in (4.3.1) is meant to estimate a nonnegative quantity, it may be replaced by $\max(|\hat{B}_k|^2-n^{-1},\ 0)$, a convention followed in the Diggle and Hall simulation study. This substitution has the practical effect of setting any estimate $\hat{B}_k$ to zero whenever its absolute value is small enough, that is, if $|\hat{B}_k|^2<n^{-1}$. Although this modification to (4.3.1) may seem trivial, in practice it is important, since the variance of $\hat{B}_k$ is large whenever $|B_k|$ is small.

A simulation study by Diggle and Hall compared cosine series density estimators where the truncation point was chosen by either the single-term stopping rule or the minimization of the cosine series equivalent of (4.3.1). The results showed that the performance of the estimator based on (4.3.1) was comparable (in terms of estimated MISE) to the estimate based on the single-term rule for some symmetric, unimodal densities and was superior for two distinctly bimodal densities. This advantage, however, only held when the single-term rule was used with small values of t, for example, $t=1$ or 2, and it vanished when t was set to 4.

Diggle and Hall also provide a proof that the cosine series analog of (4.3.1) is an asymptotically consistent estimator of the MISE in the sense that $\hat{J}/J \to 1$ as m and n increase. One practical drawback to the above stopping rule is that it requires much more extensive computation to find a truncation point than does the single-term rule. This is because many sample Fourier coefficients must be estimated and summed to calculate (4.3.1) while the single-term rule only requires the use of one estimated coefficient at a time.

4.4 Hart stopping rule

Hart (1985) has proposed a method for choosing the truncation point m of a density estimator that is similar to the procedure described in the previous section based on the determination of the value of m that minimizes an estimator of the MISE. Hart's MISE estimator can be derived by rewriting expression (4.1.7) as

$$J\left(\hat{f},f\right) = 2n^{-1}\sum_{k=1}^{m}\left(1-\left|B_k\right|^2\right) - 2\sum_{k=1}^{m}\left|B_k\right|^2 + 2\sum_{k=1}^{\infty}\left|B_k\right|^2$$
$$= \frac{2m}{n} - \frac{2(n+1)}{n}\sum_{k=1}^{m}\left|B_k\right|^2 + \sum_{k=1}^{\infty}\left|B_k\right|^2$$

and observing that the last term of the above expression does not depend on m. Hence, $J(\hat{f},f)$ is minimized by the same value of m that maximizes

$$H\left(\hat{f},f\right) = \frac{(n+1)}{n}\sum_{k=1}^{m}\left|B_k\right|^2 - \frac{m}{n}. \tag{4.4.1}$$

Using (4.4.1), an unbiased estimator of H can be constructed:

$$\hat{H}\left(\hat{f},f\right) = \frac{(n+1)}{(n-1)}\sum_{k=1}^{m}\left|\hat{B}_k\right|^2 - \frac{2m}{(n-1)}.$$

The method proposed by Hart is based on the computation of $\hat{H}(\hat{f},f)$, for all values of $m \leq M_n$, where M_n is a prespecified number dependent on the sample size n, followed by the determination of the truncation point m that maximizes $\hat{H}(\hat{f},f)$. Hart provides a proof that if m is chosen by the above method where $M_n=O(n)$, then the resulting estimator $\hat{f}(x)$ is asymptotically consistent in the sense that its integrated square error (ISE),

$$\mathrm{ISE}\left(\hat{f},f\right) = \int_0^1 \left\{\hat{f}(x)-f(x)\right\}^2 dx,$$

converges to zero in probability as $n\rightarrow\infty$.

Hart reported results from a simulation study designed to compare the above method to the single-term rule described previously for several different types of densities. For the autoregressive density—the steeply peaked, distinctly trimodal density shown in Figure 4.4—the method outlined above led to estimates that were superior in terms of estimated MISE to the single-term method. However, for two other densities, a standard normal and Chi-square (with four degrees of freedom), the situation was reversed and the single-term method consistently performed about 10–15% better than the above procedure. For a fourth density, a Beta, the two methods were quite comparable.

One weakness of the Hart method is that the choice of truncation point may be sensitive to the choice of M_n. For example, for samples of size 25, Hart specified M_n to be 20 for the normal and Chi-square densities, and 10 for the Beta density. No firm criteria for selecting M_n were recommended. Yet, as explained more fully in the next section, it is possible that the variance of $\hat{H}(\hat{f},f)$ may be inflated by an injudicious choice of M_n.

4.5 Fellner multiplier sequence

Up to this point we have only considered multiplier sequences consisting solely of ones and zeroes. Removing this constraint and allowing the multipliers to assume any real value offers the potential of much greater control in optimizing an estimator. Realizing this potential, however, is another matter, and, as will be seen below, many of the difficulties that affect stopping rules affect general multiplier sequences as well.

Thus, in this section we not only introduce a particular real-valued multiplier sequence, but also consider its relationship to the stopping rules of the previous sections.

As a starting point, consider the unweighted MISE of $\hat{f}(x)$ defined by (4.1.6). Using ordinary calculus methods, Watson (1969) showed that $J(\hat{f},f)$ is minimized by the sequence of constants

$$b_k^W = \frac{n|B_k|^2}{[(n-1)|B_k|^2+1]}, \quad k=1,2,\ldots. \tag{4.5.1}$$

These real numbers will be referred to below as the Watson multiplier sequence.

Fellner and Tarter (1971) applied a similar approach to the problem of estimation. Note that

$$\hat{J}\left(\hat{f},f\right) = \frac{2}{(n-1)}\sum_{k=1}^{\infty}\left\{b_k^2\left(1-\left|\hat{B}_k\right|^2\right)+\left(1-b_k^2\right)\left(n\left|\hat{B}_k\right|^2-1\right)\right\} \tag{4.5.2}$$

is an unbiased estimator of $J(\hat{f},f)$. The sequence that minimizes (4.5.2) is

$$\hat{b}_k^F = \frac{n-|\hat{B}_k|^{-2}}{(n-1)}, \quad k=1,2,\ldots. \tag{4.5.3}$$

Below, this will be referred to as the Fellner multiplier sequence.

Still remaining is the question of exactly how to apply the Fellner sequence to produce an estimator with good MISE characteristics. At first glance it would appear reasonable to substitute $\hat{b}_k^F$ for b_k, $k=1,2,\ldots,m$, and set $b_k=0$, $k>m$, in $\hat{f}(x)$, where m is chosen to be some large fraction of the sample size. However, experience has shown that applying the Fellner multipliers to an arbitrary number of terms in $\hat{f}$ can lead to estimates with large errors. One reason for this is that one cannot rely on the Fellner sequence to "taper" the coefficients of the raw Fourier series estimator, that is, to de-emphasize the higher-order (and smaller-magnitude) terms. Note that as $k\to\infty$, $\hat{B}_k\to 0$, but $\hat{b}_k^F$ tends to a very large negative number. To counter this tendency, Fellner and Tarter recommended setting $\hat{b}_k^F$ to zero whenever the value calculated by (4.5.3) becomes negative. This has the practical consequence of excluding from the estimator any coefficient with magnitude less than n^{-1}—a condition identical to that imposed by the Diggle and Hall stopping rule.

A second motivation for applying the Fellner sequence cautiously is related to the variability of the sample Fourier coefficients. Recall that the variance of $\hat{B}_k$ is given by

$$\mathrm{Var}(\hat{B}_k) = n^{-1}\left(1-|B_k|^2\right).$$

Thus, for fixed n, the variance of the sample coefficient $\hat{B}_k$ is large when the true Fourier coefficient B_k is small, and can reach a maximum of n^{-1} if B_k is zero. Since, as noted above, $\hat{B}_k \to 0$ as $k \to \infty$, the high-order sample coefficients tend to have larger variances than the low order coefficients. This means that the higher order Fellner multipliers are characteristically variable as well and should not be expected to taper gradually, and predictably, to zero.

Of course, the Fourier coefficients of different densities decrease to zero at different rates. For example, the coefficients of the symmetric normal distribution decay very rapidly while those of the asymmetric lognormal density decrease much more slowly. Hence, assuming that the normal and lognormal are parametrized to have equal interquartile ranges or standardized in some other way, many more coefficients of a lognormal density can be accurately estimated than can coefficients of a normal density. Naturally, for a given distribution and sample size only a finite number of Fourier coefficients can be accurately estimated and then incorporated into an estimator. As the indices of estimates of the remaining coefficients increase, the variances of these statistics will approach the limit n^{-1}.

The large variability of the high-order $\hat{B}_k$ terms has consequences not only for the application of the Fellner multipliers but also for the stopping rules discussed in the previous sections. This is because the stopping rules and the Fellner sequence both depend on an accurate estimation of the sample Fourier coefficients. For example, Hart's stopping rule relies on $\hat{H}(\hat{f},f)$, an estimator of a quantity proportional to the MISE. From (4.4.1) it can be seen that $\hat{H}(\hat{f},f)$ contains the term

$$\sum_{k=1}^{m}\left|\hat{B}_k\right|^2,$$

where m is allowed to vary over a range of positive integers, that is, $m \in \{1,\ldots,M_n\}$.

In his simulation study, Hart specified different values of M_n for the estimation of different densities, but suggested no single protocol for choosing M_n in general. It is clear that if M_n is chosen inappropriately, that is, if M_n is specified as too large a number for the particular density being estimated, $\hat{H}(\hat{f},f)$ will include some terms $\hat{B}_k$ with large variance. Thus, the variance of $\hat{H}(\hat{f},f)$ can also be inflated by the wrong choice of M_n.

A similar problem affects the Fellner multiplier sequence. High-order multipliers are dependent on high order sample Fourier coefficients, and thus, for some distributions and sample sizes, they will be unreliable. It therefore seems essential to apply the Fellner multipliers only to a carefully chosen number of terms of the Fourier series estimator.

An obvious approach to this problem is to determine the number of terms to be included in an estimator by one of the stopping rules outlined in the previous sections, and then apply the Fellner multipliers only to the included terms. The dual nature of kernel and Fourier series estimators discussed in Chapter 3 suggests that this strategy can be viewed as a two-step procedure for estimating an optimal kernel. The first step restricts the class of candidate kernel functions to those that can be represented by a Fourier expansion of m nonzero terms (actually $2m+1$ terms corresponding to the indices $-m \le k \le m$). The second step then estimates the Fourier coefficients of the best kernel (best at minimizing the estimated MISE) from within this restricted class.

Although the argument leading to this two-step procedure for estimating a density is heuristic, there is some theoretical underpinning to support its validity. Note that the MISE of an estimator which combines a stopping rule and a finite multiplier sequence, that is, an estimator where b_k, $k=1,2,\ldots,m$, are unconstrained and for $b_k=0$ for $k<m$, is

$$J\left(\hat{f},f\right) = 2\sum_{k=1}^{m}\left\{n^{-1}b_k^2\left(1-\left|B_k\right|^2\right)+(1-b_k)^2\left|B_k\right|^2\right\}+2\sum_{k=m+1}^{\infty}\left|B_k\right|^2 .$$

An unbiased estimator of this quantity is

$$\hat{J}\left(\hat{f},f\right) = \frac{2}{(n-1)}\sum_{k=1}^{m}\left\{b_k^2\left(1-\left|\hat{B}_k\right|^2\right)+\right. \tag{4.5.4}$$

$$\left.(1-b_k)^2\left(n\left|\hat{B}_k\right|^2-1\right)\right\}+\frac{2}{(n-1)}\sum_{k=m+1}^{\infty}\left(n\left|\hat{B}_k\right|^2+1\right).$$

Since the last term of (4.5.4) does not depend on $\{b_k\}$, the multipliers that minimize this estimator are a finite subset of the Fellner multipliers. This implies that no matter how the truncation point of a Fourier series estimator is chosen, the estimator can be improved by applying the Fellner multiplier sequence to its included terms, $\hat{b}_k^F, k=1,2,\ldots,m$. This conjecture was tested in a simulation study by Lock (1990), the results of which are discussed in the next section.

4.6 Estimated Watson multipliers

The Fellner multipliers discussed in the previous section were considered constants chosen to minimize an estimator of the MISE. Another approach to choosing a multiplier sequence can be based on treating the multipliers as estimators themselves. Such an approach is taken in Lock (1990), where an estimator of the optimal Watson multiplier sequence is proposed.

Finding a completely unbiased estimator of the Watson multipliers is difficult because expression (4.5.1) for b_k^W includes the term $|B_k|^2$ on both sides of a quotient. On the other hand, since $b_k^W \to 1$ as $n\to\infty$, any multiplier that converges to 1 in probability as $n\to\infty$ will be an asymptotically unbiased estimator of b_k^W. This implies that for large samples, the choice of multiplier sequence is not particularly crucial for minimizing bias. For example, $b_k=1, k=1,2,\ldots$, the multiplier sequence of the raw Fourier series estimator, is an asymptotically unbiased estimator of b_k^W.

For small sample sizes, however, the biasedness of an estimator may be more critical. The bias of the multiplier sequence $b_k=1$, $k=1,2,\ldots$, is equal to

$$\mathrm{E}(1-b_k^W) = \frac{1-|B_k|^2}{\left((n-1)|B_k|^2+1\right)},$$

a quantity that decreases to zero only as n^{-1}. The multiplier proposed by Lock, which has a form similar to that of the Watson multiplier,

$$\hat{b}_k^W = \frac{n|\hat{B}_k|^2}{[(n-1)|\hat{B}_k|^2+1]}, \quad k=1,2,\ldots,$$

has a bias component that decreases to zero as n^{-2}.

The practical difference between $\hat{b}_k^W$ and $\hat{b}_k^F$ can be made clear by rewriting $\hat{b}_k^W$:

$$\hat{b}_k^W = \frac{n}{[(n-1)+|\hat{B}_k|^{-2}]}.$$

Although both multipliers contain the term $|\hat{B}_k|^{-2}$, it appears in the numerator of $\hat{b}_k^F$ and in the denominator of $\hat{b}_k^W$. This means that Fourier coefficients of small magnitude will tend to be emphasized in $\hat{b}_k^F$ and de-emphasized in $\hat{b}_k^W$. Since small magnitude coefficients are the hardest to estimate accurately, the Fellner multiplier has a tendency to emphasize the most unreliable terms of the estimator. This is why it is advisable to set $\hat{b}_k^F$ equal to zero whenever $\hat{B}_k$ is small. On the other hand, the estimated Watson multiplier is bounded within the interval [0,1], and as $\hat{B}_k \rightarrow 0$, $\hat{b}_k^W \rightarrow 0$. Thus, $\hat{b}_k^W$ possesses the desirable property of gradually tapering coefficients of the Fourier series estimator to zero as $k \rightarrow \infty$.

In a simulation study reported by Lock (1990), each of the stopping rules described in the earlier sections were combined with the estimated Watson multipliers in a two-step density estimation procedure. Specifically, a stopping rule was used to first determine the total number of terms m to include in the estimator, and second, a general multiplier sequence was determined and applied to only the first m terms of the estimator. The results showed that regardless of distribution, sample size, or stopping rule, the application of $\hat{b}_k^W$ led to a decrease in the estimated MISE compared to the raw estimator.

The simulation results also lend some support to the assertion made earlier in this section: the choice of multiplier sequence is less critical for larger sample sizes. In general, the degree to which the addition of $\hat{b}_k^W$ improved the raw estimates decreased as sample size increased.

4.7 The weighted MISE

The stopping rules and multipliers discussed in the previous section were all designed to be optimal in an overall sense by emphasizing the unweighted MISE. But as mentioned in Chapter 3, an overall view is sometimes not the most useful one for a curve estimation researcher. In these situations, the weighted MISE can be employed to improve the performance of the estimator over certain subregions of the density's support. Specifically, a "weighted" estimator can be produced that trades accuracy in one region of the estimate for increased resolution in another, possibly more important, region.

An example of a potentially useful weight function is

$$w_R(x) = 1 - \alpha \sin(2\pi x), \quad x \in [0,1], \quad \alpha \in [0,1].$$

(Limiting the value of α to within [0,1] ensures that $w_R(x)$ will be nonnegative.) Figure 4.5 is a graph of $w_R(x)$ over the interval [0,1] for

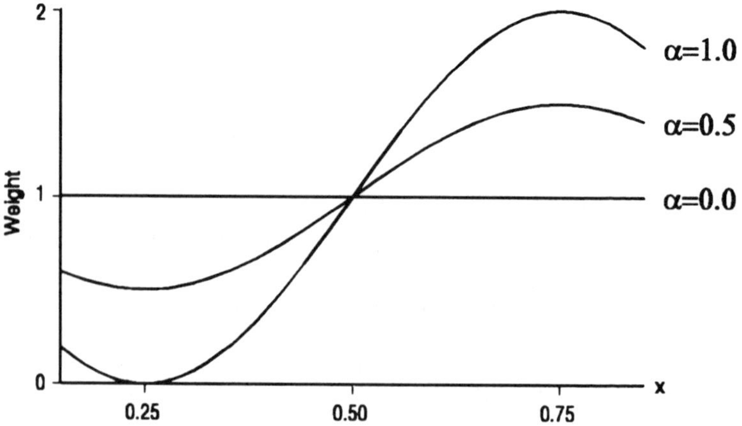

Figure 4.5. *Graph of the weight function $w_a(x)$ for three different values of the parameter α.*

three different values of the parameter α. Note that at $\alpha=0$, $w_R(x)$ is a horizontal line, the uniform weight function of the unweighted MISE. The curves corresponding to the parameters $\alpha=.5$ and $\alpha=1$ both attain their maxima at $x=3/4$, the latter at one and a half times the height of the former. Thus, for a density whose support is on $[1/4, 3/4]$ or which has been transformed to lie within $[1/4, 3/4]$, $w_R(x)$ can be used to emphasize the right tail of a density at the expense of the left side. Such a tradeoff may seem particularly worthwhile to a researcher interested in comparing the long-term, that is, right tail, survival characteristics of different populations.

The choice of α controls how much overall accuracy the researcher is willing to sacrifice for improvement in the right tail, from a minimum at $\alpha=0$ to a maximum at $\alpha=1$. Correspondingly, the value $\alpha=0$ weights all subregions equally, while the choice of $\alpha=1$ emphasizes the estimation of the right tail to the exclusion of all other regions of the curve.

The choice of the weight function

$$w_L(x) = 1+\alpha \sin(2\pi x), \quad x \in [0,1], \quad \alpha \in [0,1].$$

will lead to an estimator most accurate in the left region of its support. As shown in Figure 4.6, the relationship between and the degree of left-tail emphasis is analogous to the corresponding relationship in $w_R(x)$.

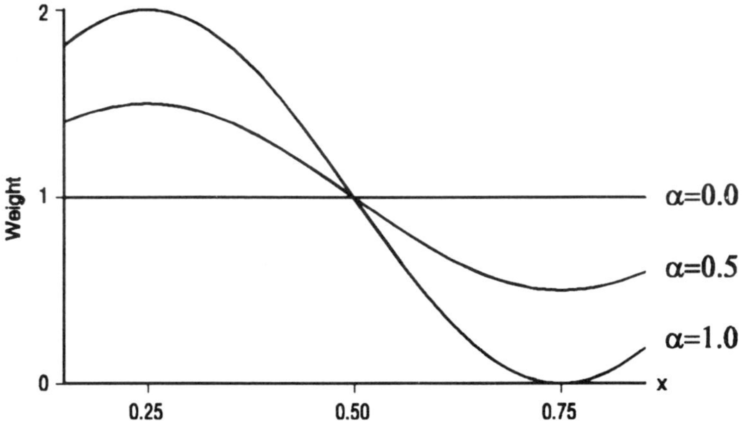

Figure 4.6. *Graph of the weight function $w_i(x)$ for three different values of the parameter α.*

The weight $w_L(x)$ might prove useful to a researcher investigating the response to very low dosages of some agent.

Given that the methods of the previous sections can produce estimates with small MISE, it is appropriate to question whether a weighted estimator could yield any appreciable improvement over the unweighted variety. Some indication that the answer to this question may be "yes" can be found through a careful examination of some of the characteristics of the MISE.

The unweighted MISE could be more precisely called a uniformly weighted error metric since individual errors, $\hat{f}(x)-f(x)$, are considered equally important regardless of where they occur. That is, an error of .01 in the left tail of an estimate contributes as much to the overall MISE as an error of .01 at the mode or in the right tail. But consider, for example, the standard normal density. The height of this density at the mode is .3989, and thus an error of .01 would be a discrepancy of less than 3%. In the tails of the distribution, however, an error of the same absolute magnitude results in a much greater relative discrepancy. For example, at $x=2.0$, the height of the standard normal density is .0540; therefore, an error of .01 is actually an error of more than 18%.

By weighting errors uniformly with respect to x, the unweighted MISE actually weights the size of the relative error at x in proportion to the square of the height of the unknown density at x. Specifically, define the relative error of $\hat{f}$ at x to be

$$\phi(x) = \frac{\hat{f}(x)-f(x)}{f(x)}.$$

Then by a simple rearrangement, the unweighted MISE can be written in terms of the relative error:

$$J(\hat{f},f,1) = \mathrm{E}\int_0^1 \left\{\hat{f}(x)-f(x)\right\}^2 \left\{\frac{f(x)}{f(x)}\right\}^2 dx \tag{4.7.1}$$

$$= \mathrm{E}\int_0^1 \{\phi(x)\}^2\{f(x)\}^2 dx.$$

Thus, the unweighted MISE actually gives more weight to relative errors in the areas where the unknown density is highest. This leads to estimates that are most accurate, in terms of relative error, near the maximum of the distribution, that is, near the mode. Since in many cases the mode of a distribution is near the center of its support, this characteristic of the unweighted MISE is sometimes called a tendency toward center-weighting.

This tendency suggests that there may be an opportunity to improve unweighted estimates, especially when regions of the density other than the mode are of most interest. For example, by a manipulation similar to that used to obtain (4.7.1), it can be shown that the weighted MISE $w(x)=[f(x)]^{-2}$ with gives uniform emphasis to all relative errors along the x axis. Methods that consider similar weight functions, as well as more general weighting procedures, will be the subject of the next sections.

4.8 Transformations and the weighted MISE

One approach to implementing a weighted density estimator is based on combining a transformation of the data with the optimization methods outlined earlier in this chapter. The basic strategy involves three steps. First the data are transformed in such a way that the density of the transformed variate is more conducive to effective curve estimation than is the pretransformed variate. Second, the density of the transformed data is estimated using a procedure based on the unweighted MISE. If a nonlinear transformation has been chosen, the data will be spread along the axis in a different manner than in the original sample, and correspondingly, the weight applied by the metric to various points will be different as well. Third, the estimate is transformed back into the space of the original data.

To help establish the relationship between the choice of data transformation and the weight function of the MISE, the following method for generalizing the orthogonal sequence of functions

$$\Psi_k(x) = \exp\{2\pi ikx\}, \quad k=0,\pm1,\pm2,\ldots, \tag{4.8.1}$$

into a new sequence of functions is useful. Let G be a strictly increasing, one-to-one function, $G:[a,b]\rightarrow[0,1]$, with a continuous first derivative $g(x)=dG(x)/dx$. Then the sequence of functions $\Psi_k(G(x))$, $k=0,\pm1,\pm2,\ldots$, is orthonormal on $[a,b]$ with respect to the weight $g(x)$, that is,

$$\int_a^b \Psi_j(G(x))\Psi_{-k}(G(x))g(x)dx = \delta_{j,k}. \tag{4.8.2}$$

This relationship can be proven by using the change of variables $y=G(x)$ as follows:

$$\begin{aligned}\int_a^b \Psi_j(G(x))\Psi_{-k}(G(x))g(x)dx &= \int_a^b \exp\{2\pi i(j-k)G(x)\}g(x)dx \\ &= \int_0^1 \exp\{2\pi i(j-k)y\}dy = \delta_{j,k}.\end{aligned}$$

Hence, by specifying different functions for $G(x)$ a wide variety of orthogonal sequences can be devised. In particular, sequences orthogonal on any finite interval of the real line can be constructed in this fashion.

Because (4.8.2) is a complete orthonormal basis a large class of densities can be represented by the series:

$$f(x) = g(x)\sum_k \beta_k \Psi_k(G(x)).$$

Analogous to the approach taken in Section 3.2, an estimator of f can be defined:

$$\hat{f}(x) = g(x)\sum_k b_k\hat{\beta}_k\Psi_k(G(x)), \tag{4.8.3}$$

where

$$\hat{\beta}_k = n^{-1}\sum_{j=1}^{n} \Psi_{-k}\left(G(X_j)\right),\ k=0,\pm 1,\pm 2\ldots. \tag{4.8.4}$$

Consider the weighted MISE of (4.8.3) where the weight function is specified as $w(x)=1/g(x)$:

$$J\left(\hat{f},f,g^{-1}\right) = \mathrm{E}\int_a^b \left\{f(x)-\hat{f}(x)\right\}^2 [g(x)]^{-1}dx$$

$$= \mathrm{E}\int_a^b \left\{ \sum_k b_k \hat{\beta}_k \Psi_k(G(x)) - \sum_k \beta_k \Psi_k(G(x)) \right\}^2 g(x)dx$$

$$= 2\sum_{k=1}^{\infty} \left\{ n^{-1} b_k^2 \left(1 - \left|\beta_k\right|^2\right) + (1-b_k)^2 \left|\beta_k\right|^2 \right\}.$$

Note that the last quantity on the right side above is almost identical to Equation (4.1.6) for the unweighted MISE of an estimator; the only difference is in the definition of the Fourier coefficients. Since the unweighted optimization methods described earlier were designed to minimize (4.1.6), it is clear that the application of any of these methods to an estimator of the form (4.8.3) would optimize it with respect to the weighted MISE where $w(x)=1/g(x)$.

As an example, consider the transformation $G(x)=(x-a)/r$, where and $a \le \min\{X_1,\ldots,X_n\}$ and $r > \max\{X_1,\ldots,X_n\}-a$, that was used in Chapter 3 to scale the sample data to lie within the unit interval. Using (4.8.3) and (4.8.4), we can construct the actual form of an estimator that includes this transformation as an integral part of its representation:

$$\hat{f}(x) = r^{-1} \sum_k b_k \hat{\beta}_k \Psi_k((x-a)/r),$$

where $\hat{\beta}_k = n^{-1} \sum_{j=1}^{n} \Psi_{-k}\left((X_j - a)/r\right)$. An algorithm for calculating an estimate which incorporates this transformation is given by Tarter, Freeman and Hopkins (1986).

It is interesting to note that although this procedure was described as a data transformation, the previous discussion makes it clear that it is equivalent to modifying orthogonal series (4.8.1). Also note that the methods described earlier in the chapter can be used to optimize this estimator with respect to the weighted MISE, where $w(x)=r$:

$$J\left(\hat{f}, f, r\right) = \mathrm{E}\int_0^1 \left\{ f(x) - \hat{f}(x) \right\}^2 r dx.$$

Since the weight function r is independent of x, $J(\hat{f},f,r)$ is seen to be a uniformly weighted metric, like the unweighted MISE. Thus, the uniform emphasis of the unweighted MISE is preserved under linear transformations of the data.

Conversely, several authors have taken advantage of the fact that a nonlinear transformation will change the emphasis of the weighted MISE metric. For example, Fellner (1974), Brunk (1978), and Anderson and de Figueiredo (1980) have each proposed procedures that rely on the same choice of transformation, an approximation to the probability integral transformation. Each of these methods involves a first step where an initial or "pilot" estimate, $\tilde{f}$, of the original data is made. The pilot curve is not required to be highly accurate since it is only the first step of a multistep process. The next step is to apply the transformation

$$Z_j = \int_0^{X_i} \tilde{f}(x)dx, j=1,\ldots,n$$

to each of the data points, X_j. If $\tilde{f}(x) \equiv f(x)$, this would be the probability integral transformation. Finally, the density of the transformed variate, Z_j, $j=1,\ldots,n$, is estimated and (4.8.3) and (4.8.4) applied. Any of the unweighted stopping rules or multipliers described earlier will optimize this estimator in terms of the weighted MISE $J(\hat{f},f,\tilde{f}^{-1})$. Thus, the procedures of Fellner, Brunk, and Anderson and de Figueiredo are very similar to procedures for minimizing the relative error of a density estimate.

Another approach that incorporates a data transformation into a kernel density estimator has been proposed by Wand, Marron and Ruppert (1991). This method also relies on a pilot estimate. However, rather than employing the pilot directly, in the form of a transformation, $\tilde{f}$ is used to choose a transformation from a family of functions in a manner analogous to the procedures described in Chapter 7. The family investigated by Wand, Marron and Rupperts involved transformations:

$$G_\alpha(x) = \begin{cases} x^\alpha \text{sign}(\alpha), & \alpha \neq 0 \\ \ln(x), & \alpha = 0 \end{cases}$$

where $x>0$. Note that this choice is an excellent one for transforming data that are highly skewed to the right since for $0<\alpha<1$, $G_\alpha(x)$, is a concave function. In a trial using data simulated from a highly skewed lognormal distribution, Wand, Marron and Rupperts showed that their automatic method for choosing a transformation led to vastly superior estimates in terms of estimated MISE.

These results show that transformation techniques are not only useful for increasing accuracy over prespecified subregions, but, in some instances, can even improve estimators in an overall sense. This is because certain densities, such as symmetric, unimodal functions like the normal curve, are estimated more efficiently than others. Generally speaking, transformations can be exploited to produce more accurate estimates overall by reducing a problem to a previously solved problem.

4.9 Weighted multiplier sequences

There is one major drawback to the use of transformation methods. To obtain weighted estimators a new set of sample Fourier coefficients needs to be calculated from the data for every different weight function used. From an examination of the algorithms published in Tarter, Freeman and Hopkins (1986), it is clear that the routine for estimating the coefficients requires far more computational effort than any other aspect of the density estimation process. Specifically, a total of Mn complex operations are required to calculate M sample coefficients from n data points, while, for example, only a maximum of M much simpler operations are needed to choose the single-term truncation point. Thus, for every different weight function considered, the transformation approach requires reiteration of the most computationally intensive routine in the density estimation process.

The method of Tarter, Freeman and Polissar (1990) circumvents this disadvantage by following a more direct approach, which retains the original set of Fourier coefficients while employing different multiplier sequences to optimize the estimator with respect to different weight functions. This method assumes the weight function is expressible by a Fourier expansion.

Assume that the weight function is expressible by a Fourier expansion:

$$w(x) = \sum_k w_k \exp\{2\pi ikx\}. \tag{4.9.1}$$

Incorporating (4.9.1) into (4.0.1), the weighted MISE can be expressed

$$J(\hat{f},f,w) = \mathrm{E}\int_0^1 \left| \sum_k \left(b_k\hat{B}_k - B_k\right) e^{2\pi ikx} \right|^2 \sum_v w_v e^{2\pi ivx} dx$$

$$= \mathrm{E}\int_0^1 \sum_k \sum_j \sum_v \left(b_k\hat{B}_k - B_k\right)\left(b_{-j}\hat{B}_{-j} - B_{-j}\right) w_v e^{2\pi i(k-j+v)x} dx$$

$$= \mathrm{E}\sum_k \sum_j w_{j-k}\left\{b_k b_{-j}\hat{B}_k\hat{B}_{-j} - b_k\hat{B}_k\hat{B}_{-j} - b_{-j}B_kB_{-j} + B_kB_{-j}\right\}.$$

Using identity (4.1.3) for the expected value of $\hat{B}_k\hat{B}_j^* \equiv \hat{B}_k\hat{B}_{-j}$, and restricting b_k to be equal to b_{-k} leads to

$$J(\hat{f},f,w) = \sum_k \sum_j w_{j-k}\left\{n^{-1}b_k b_j\left(B_{k-j} - B_kB_j\right) + \left(1-b_k\right)\left(1-b_j\right)B_kB_{-j}\right\}. \tag{4.9.2}$$

Following the same general approach used to derive the Fellner multipliers described in Section 4.5, the authors constructed an unbiased estimator of (4.9.2) and then used calculus and linear algebra methods to find the multiplier sequence that minimized this estimator.

Tarter, Freeman and Polissar tested this method in an experiment involving many varieties of simulated survival data. In order to emphasize the estimation of long-term survival probabilities, they chose the right tail-enhancing function $w_R(x)$ shown in Figure 4.5. The resulting estimated survival curves proved to be superior to either life table or Kaplan-Meier (1958) estimates over the right tail of the distribution even when judged by criteria different from those used to construct the estimators. As will be detailed in Chapter 9, this was found to be the case even when the degree of abridgement of the life table was chosen to be an optimal user control setting, even though comparable settings of the new methods were estimated values, not optimal values.

Of course, like any other orthogonal series estimator, the weighted density estimator described by Tarter, Freeman and Polissar can only include a finite number of sample coefficients. This leads to the question of how to choose those terms to be included in the estimator, and consequently, how many weighted multipliers will be required. One reasonable solution would be to use one of the stopping rules described earlier. These criteria, however, are all designed to select a truncation point that minimizes the unweighted MISE. Given the objective here—to construct estimators that minimize the weighted MISE—it is logical to assume the performance of an estimator can be most improved by choosing not only a multiplier sequence, but also a truncation point that minimizes the weighted MISE.

This approach is taken by Lock (1990) where a weighted stopping rule is proposed that follows a line of reasoning similar to the unweighted single-term rule described in Section 3.2. Essentially, the rule relies on estimating the change in the weighted MISE due to the inclusion of an additional term to the raw estimator. If the estimated change is negative, then the inclusion of that term is warranted.

To increase the computational efficiency of the estimation algorithms, Lock restricted the weight function to be from a very simple family of functions: those expressible by a three-term Fourier expansion. This includes the two weight functions, $w_R(x)$ and $w_L(x)$, described in Section 4.7. For example, the coefficients of the right-tail weighting function $w_R(x)$ are $w_0=1$, $w_1=w_{-1}=i\alpha/2$, and $w_K=0$, $k\neq 0,\pm 1$. Under these circumstances, the change in $J(\hat{f},f,w)$ due to the addition of the sth term (to an estimator which already contains $s-1$ coefficients) can be estimated by

$$\Delta\hat{J}_s = \frac{2w_0}{(n-1)}\left(1-\left|\hat{B}_s\right|^2\right)-\frac{2w_0}{(n-1)}\left(n\left|\hat{B}_s\right|^2-1\right) + \frac{4}{(n-1)}\mathrm{R}\left\{w_{-1}(\hat{B}_1-\hat{B}_s\hat{B}^*_{s-1})\right\} + \frac{4}{(n-1)}\mathrm{R}\left\{w_{-1}(n\hat{B}_{s+1}\hat{B}^*_s-\hat{B}_1)\right\}, \tag{4.9.3}$$

where R$\{z\}$ is the real part of the complex number z. To minimize the weighted MISE, the sth term should be included in $\hat{f}(x)$ only if (4.9.3) is negative. This is equivalent to the inclusion test: Include the sth term only if

$$w_0\left(2-(n+1)\left|\hat{B}_s\right|^2\right)+2\mathrm{R}\left\{w_{-1}(\hat{B}_1-\hat{B}_s\hat{B}^*_{s-1})\right\}-2\mathrm{R}\left\{w_{-1}(n\hat{B}_{s+1}\hat{B}^*_s-\hat{B}_1)\right\}$$

is less than zero. Note that when $w_0=1$ and $w_1=w_{-1}=0$, the Fourier coefficients of the uniform weight function, the above criterion is exactly equivalent to the single-term inclusion test.

To implement the weighted stopping rule, the inclusion test is applied to the terms $\hat{B}_k$ sequentially. After t consecutive terms have failed to satisfy the inclusion criterion, the testing of terms is halted. The truncation point is then set to m, the index of the highest-order term which passed the test. In a simulation study, Lock showed that a combination of weighted stopping rule and the weighted multiplier sequence of Tarter, Freeman and Polissar (1990) was more effective at emphasizing certain subregions of a density than either method alone and more effective than using an unweighted optimization technique. This effect was most pronounced when estimating highly skewed, long-tailed distributions such as the lognormal or Weibull functions.

There are many applications of the use of weighted multiplier sequences. For example, a very common statistical problem is the decomposition of mixture components which are separated by distinctly valued location parameters. In the two-component case this sort of problem can be characterized as involving right-most and left-most components. Consequently, as we will see in the next chapter, it is critical to estimate each isolated component differentially, in the sense that a weight function w that emphasizes the left side of the support region will increase resolution of the left-most component and vice versa.

CHAPTER 5

Mixture decomposition applications

In this chapter we will examine methodology for analyzing finite mixture distributions that are composites of two or more statistical models. Specifically, a mixture is thought to consist of c subpopulations within which variates are homogeneously valued and between which variates are heterogeneously valued. Information concerning subgroup membership is available only through the variate values themselves. Also, although most of this chapter will deal with methodology that does not depend on prior knowledge of c, specific methods that do depend on this knowledge will be reviewed.

5.1 Mixtures and multipliers

The simplest case of the finite mixture model is based on the assumption that it is useful to individualize the subcomponents f_1 and f_2 of the mixture density

$$f(x|p,\mu_1,\mu_2,\sigma) = \{pf_1([x-\mu_1]/\sigma) + (1-p)f_2([x-\mu_2]/\sigma)\}/\sigma,$$

parametrized by p, μ_1, μ_2, and σ. The lambda multipliers provide an elegant way of separating mixture components and are easy to implement using series forms of curve estimation.

Before dealing with these multipliers it is informative to introduce the Fejér multiplier sequence, which was the first of the large number of multiplier systems used for specialized mathematical purposes. The Fejér sequence contains equally spaced evaluations of a kernel transform (KT) that has a triangular shape, while the lambda multipliers correspond to a KT that is similar in shape to either the normal pdf or the normal's reciprocal. The normally shaped KT corresponds to lambda multipliers that decrease contrast and the reciprocal normally shaped KT corresponds to lambda multipliers that increase contrast. Here, the term contrast refers to the slurring together of mixture components—contrast reduction—or the separation of components—contrast enhancement.

As will be shown in Section 5.4, the decomposition of certain mixtures by the lambda KT does not require a priori knowledge of the number of individual components that constitute the mixture. This is one distinct difference between the lambda method and cluster analysis. Generally speaking, the lambda KT allows mixture subcomponent separation within the frequency or Fourier domain, while cluster analysis methodology separates data points.

The following example details the steps by which many models of the form $f(x|p,\mu_1,\mu_2,\sigma) = \{pf_1([x-\mu_1]/\sigma) + (1-p)f_2([x-\mu_2]/\sigma)\}/\sigma$, can be parsed by the lambda KT. Naturally, there are mathematical conditions on the functions f_1 and f_2 which ensure that the lambda KT will reduce contrast. These will be discussed in Sections 5.5 and 5.6.

Here $n=1000$ iid variates were generated from $f(x|0.4,0,3.3,1)$, where $f_1(x)\equiv f_2(x)\equiv N[x|0,1]$, which yielded the solid curve shown in Figure 5.1. The dotted curve shown in this figure was obtained by applying a lambda multiplier sequence $\{b_k^{(\lambda)}=\exp[2\pi^2k^2\lambda^2]\}$ with the user control setting $\lambda=0.8$. This setting was selected so that as $n\to\infty$, the estimator would approach the density $f(x|0.4,0,3.3,0.6)$. Owing to the

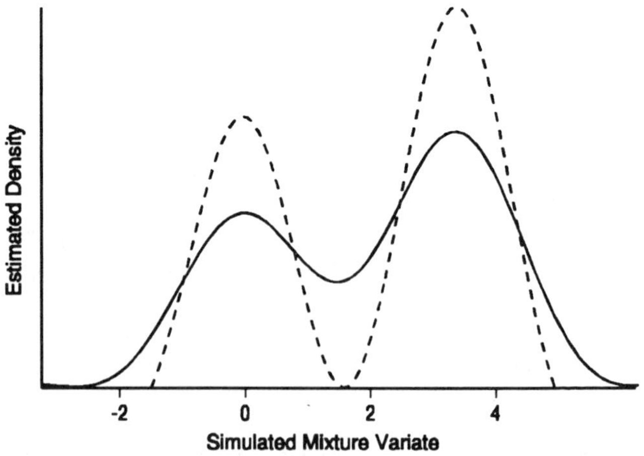

Figure 5.1. *Density estimate where the λ method has been used to reduce component overlap to zero. For the dashed curve here, $\lambda=0.8$. The data are simulated from a mixture of normal densities.*

limitations of the estimation process, it is not advisable to use larger values of λ. Thus, an artificial estimator was constructed, which in the limiting case has all properties but one identical to those of the original estimator. The one important exception is that component overlap was reduced to zero at some point, referred to as x_0.

Once there was no longer any overlap, a second multiplier procedure was used to mask, blank or zero out the curve to one side of x_0 based on a simple and well-known linear relationship for the Fourier coefficients of the product of two functions. Here, one of the two functions is the overlap-reduced curve and the other is the indicator function of the interval L over which the masking is to take place. Specifically, this second function is $I_L(x)$, defined to equal one if x is within L and zero elsewhere. Once the masking multipliers were put into place, the process that had reduced the component overlap was reversed. For example, since a positive value, $\lambda=0.8$, had been used to create the dashed curve shown in Figure 5.1, after masking out the curve to the left side of x_0, a negative value $\lambda=-0.8$ was used to construct the curve shown in Figure 5.2. By integrating this curve over the interval $[x_0,1]$, an estimated value of 0.5962 was obtained for the true mixing parameter $(1-p)=0.6$.

Figure 5.3 shows both a normal and logistic curve fitted to the mixture component that was individualized as shown above. Here the normal

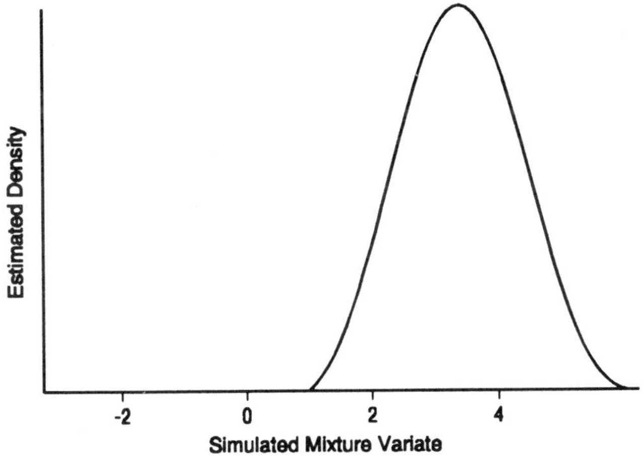

Figure 5.2. *Estimated right-most component of density shown in Figure 5.1. This component was estimated by first isolating the right-most portion of the variance-reduced density in Figure 5.1 and the reversing the λ procedure.*

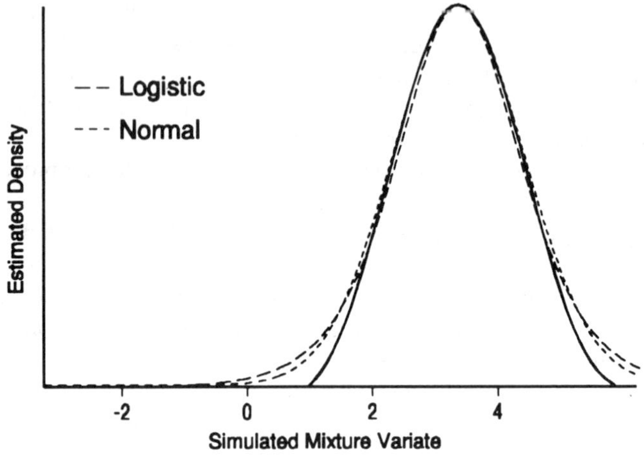

Figure 5.3. *Logistic and normal density functions fitted to the estimated right-component density shown in Figure 5.2.*

(the curve drawn with dashes) provides a slightly better fit than the logistic (the curve drawn with dots) to the empirical estimator (the solid curve).

Chapter 7 will describe techniques that apply specific Fourier coefficient estimators to obtain estimators of an assumed-elementary-function-model's parameters. If a researcher is uncertain about a mixture component's representation by a model, the methodology discussed in the next chapter can be applied. One of the advantages of this selection approach stems from the use of information from the midsection, as opposed to the density tail regions. Mixture decomposition procedures characteristically yield poor tail estimators of individualized curves over regions where components overlap. For this reason, it is preferable to supplement the simple model fitting approach shown in the previous figure with the display of the curve, $R(x)$, introduced in Section 1.1; here estimated over the support midsection. The estimated $R(x)$ associated with the normal curve shown by the solid line in Figure 5.4 closely fits a horizontal line while the $R(x)$ curves associated with logistic, extreme value and Cauchy models shown in Figure 5.5 are *not* horizontal lines. Because the estimated normal-based $R(x)$ is slightly, but perceptively, straighter than the estimated logistic-based and other $R(x)$ curves, reasoning outlined in the next two chapters leads to the conclusion that the normal should be judged the best choice of underlying model.

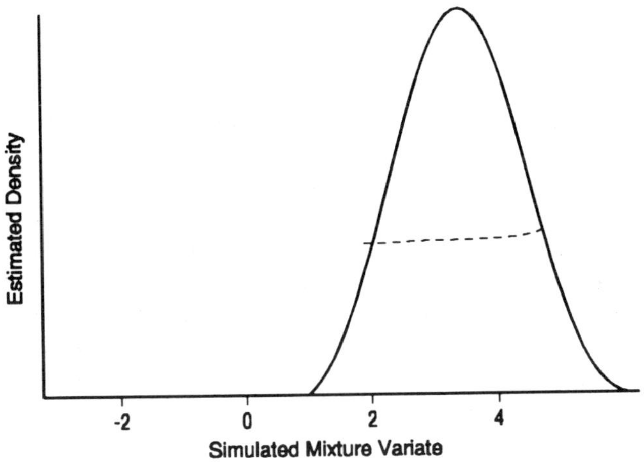

Figure 5.4. *Normal-based R(x) function superimposed upon the estimated density of the right-most component.*

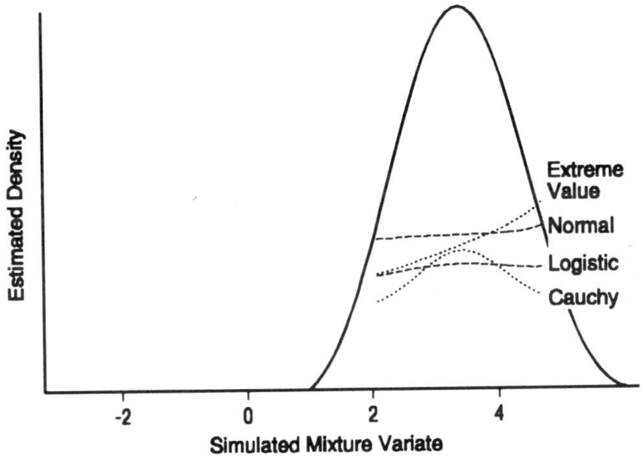

Figure 5.5. *Normal, Logistic, Cauchy and Extreme Value R(x) functions superimposed on the estimated density shown in Figure 5.2. Note the Extreme Value and Weibull densities are related by the log transformation.*

The estimated mode was 3.34, and $R(3.34)=1.039$, which, as will be shown in the next chapter, provides an estimator of the model scale parameter. The true mode and scale parameter of the component were 3.3 and 1, respectively. Thus, as is typical for a sample of size $n=1000$ from a population whose datum density is $f(x|0.4,0,3.3,1)$, it was possible to accurately estimate the mixing parameter, location and scale parameters of this mixture component. When the above steps were repeated after blanking out the larger component, the estimate of the mode (whose true value was zero) of the smaller component was determined to be 0.0156, while $R(0.0156)$ was found to equal 0.9565. (The true value of this component's scale parameter was 1.) Although these parameter estimators were accurate, it was difficult to decide whether a normal or a logistic was the more appropriate choice of model for the $p=.4$ component, even though upon detailed examination of the $p=.6$ component, this distinction can be made by using the two $R(x)$ curves shown in Figure 5.6.

5.2 Approaches to mixture decomposition

A substantial body of literature deals with graphical decomposition procedures that depend on the properties rather than the parametric

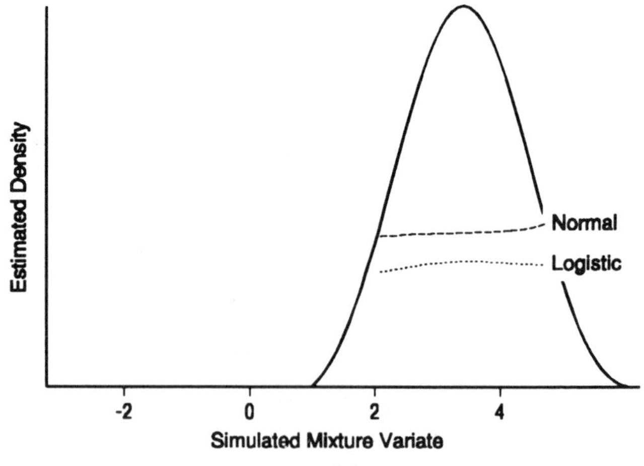

Figure 5.6. *Normal and Logistic R(x) functions superimposed on the estimated density of the right-most component. The slightly straighter line of the Normal R(x) suggests that the estimated density is better fit by the Normal model than the Logistic.*

structure of mixture models. For example, Buchanan-Wollaston and Hodgeson (1929) used the estimated distribution modes to decompose mixtures of normals. Eisenberger (1964) used both mode and trough location (a point at which an estimated curve attains a local minimum) for this same purpose. Weichelberger (1961) and Bhattacharya (1967) used graphs of the ratios between estimated density evaluations. Weichelberger's method uses ratios of sample frequencies from adjacent equal length class intervals to approximate a straight line plotted on semilogarithmic paper.

A method like Weichelberger's can be implemented using modern curve estimation approaches. Suppose there is substantial separation between the two components of $f(x|p,\mu_1,\mu_2,\sigma)$, $pf_1([x-\mu_1]/\sigma)/\sigma$, $(1-p)f_2([x-\mu_2]/\sigma)/\sigma$, where f_1 and f_2 are standard normal, $\mu_1<\mu_2$ and $|\mu_1-\mu_2|>\sigma$. Then, $f(x|p,\mu_1,\mu_2,\sigma)$ is approximately proportional to

$$\frac{p}{\sqrt{2\pi}\sigma}\exp\left\{-\frac{1}{2}\left(\frac{x-\mu_1}{\sigma}\right)^2\right\}$$

over the left-tail of its support region. Thus, the derivative of $\log f(x|p,\mu_1,\mu_2,\sigma)$ is approximately equal to $-\sigma^{-2}[x-\mu_1]$ within this same region, and estimates of this derivative can be used to determine μ_1 and σ graphically.

There is a major drawback to the Weichelberger approach to mixture decomposition. Weichelberger's derivative of the log density estimator only estimates $y=-\sigma^{-2}[x-\mu_1]$ in a manner that is functionally independent of the subset of mixture model parameters, $\{p$ and $\mu_2\}$, when it can be assumed that estimates over tail regions are accurate. This assumption is particularly suspect when there is substantial overlap. If overlap exists, which is likely to be the case when $|\mu_1-\mu_2|$ is not appreciably larger than σ, a very large sample would be needed to obtain reliable estimates far out along the estimated density tail.

The assumption that the tail region overlap of mixture components is inappreciable has been used by Harding (1949), Preston (1953) and Cassie (1954) to devise graphical techniques for the estimation of left-most or right-most component location parameters. The idea that numerical or graphical subtraction procedures could be used to sequentially deal with mixture decomposition was first discussed by Szigeti (1947) and Labhart (1947), while Noble, Hayes and Eden (1959) compared estimated densities by overlaying cathode ray tube displays with curves printed on transparent paper.

As far as nongraphical approaches are concerned, Cicchinelli (1962), Subrahámian (1963), and Hasselblad (1966) describe new likelihood procedures, while Frankowski (1967) constructed moment-based procedures for mixture decomposition based, in part, on an in-depth survey of previous methods, while Izenman and Sommer (1988) reviewed and applied kernel-based procedures.

Although his work did not directly deal with either data analysis or computational considerations, Medgyessy (1961) brought the lambda method, first proposed by the mathematician Doetsch (1928 and 1936) to the attention of the applied statistical community. However, Doetsch's papers, Medgyessy's 1961 book, and a subsequent paper by Gregor (1969) made no distinction between: (1) the problem of decomposing a mixture where the actual function could be directly evaluated at an equally spaced sequence of points and, (2) the problem of mixture decomposition where data are available in the traditional statistical sense, as input from an iid sample.

The problem of applying Doetsch's Fourier-based method to sample data was first considered by Kronmal (1964). Two papers based on Kronmal's approach were published by Stanat (1966 and 1968). The advantages of applying this technique in the bivariate case were also discussed by Tarter and Silvers (1975), Titterington, Smith and Makov (1985, Section 7.7.4), while Tarter, Rigsbee and Wong (1976) and Tarter (1979b) introduce the specific techniques used to obtain Figures 5.1 through 5.9.

For $c>1$ the lambda KT's applicability to the c-component mixture model $f_c(x|p_1,p_2,\ldots,pc_{-1},\mu_1,\mu_2,\ldots,\mu_c,\sigma_1,\sigma_2,\ldots,\sigma_c)$ defined as

$$p_1f_1([x-\mu_1]/\sigma_1)/\sigma_1+p_2f_2([x-\mu_2]/\sigma_2)/\sigma_2+\ldots$$
$$+(1-\sum_{j=1}^{c-1}p_j)f_c([x-\mu_c]/\sigma_c)/\sigma_c$$

depends on two classes of assumptions. The first is that unimodal $f_1,f_2,\ldots$ are symmetric about a mode with coordinate zero. The second is that the parameters $\sigma_1^2,\sigma_2^2,\ldots,\sigma_c^2$ are individually proportional or have some other monotonic relationship to the population variances (second cumulants) of the respective subcomponents.

Under these assumptions, the parameters $\sigma_1^2,\sigma_2^2,\ldots,\sigma_c^2$ of the mixture can be altered by specific lambda KTs in a manner that reduces subcomponent overlap. Even though the value of c may not actually be known prior to use of the lambda KT, the structure of the mixture model (specifically, the addition operation that separates mixture components) ensures that a single multiplier sequence will distribute among c-components. Like a prescribed drug that is taken orally or injected into the bloodstream, the lambda KT is designed to target precisely those portions of individual mixture components, the parameters $\sigma_1^2, \sigma_2^2, \ldots, \sigma_c^2$, which, when reduced, render mixture decomposition manageable. Like many prescribed drugs this process does not work satisfactorily in all situations. For example, it cannot directly deal with mixtures of components that have identical location parameters but different scale parameters, since in this situation even though the common value of all component location parameters can be changed, this change will not help separate mixture components.

As will be shown in Section 5.5, modifications are required for application to mixtures of densities like the Cauchy, which have a scale parameter that cannot be associated with the second cumulant. Even with this and other modifications the methodology described below has many limitations. Nevertheless, with the exception of the procedure described by Jones (1991), the lambda KT is the one tool available today that can systematically reverse the bloating or excess smoothing, $\overline{\sigma}^2$, which can be a misleading artifact produced by many curve estimation procedures. By making certain mixtures appear more homogeneous than they actually are, this artifact has the pernicious effect of masking the requirement that a mixture extended model be assumed in place of a simple model. We will now derive a basic formula that quantifies $\overline{\sigma}^2$.

5.3 Fejér multipliers as intermediates between Dirichlet and lambda multipliers

Although there are several ways to describe the details of the lambda decomposition procedure, it is easiest to base this description on the kernel transform (KT). The multipliers $\{b_k^{(\lambda)}\}$, which are equally spaced evaluations of the lambda KT, resemble a Fejér sequence of multipliers. The Fejér sequence was the first of many multiplier systems devised for theoretical mathematical purposes. Hence, for historical and other reasons, we will now describe the derivation and application of this system.

Suppose that a continuous variate's pdf is either some non-L_2 function, or some discontinuous function, like the rectangular function $I_{[\omega_1,\omega_2]}(x)/[\omega_2-\omega_1]$, $0<\omega_1<\omega_2<1$. The Fejér multiplier system can then be applied to obviate all theoretical barriers to use of the $\{\exp(2\pi ikx)\}$ orthogonal sequence as a basis for the representation of a continuous variate's pdf. However, before turning to the representation of a continuous variate pdf, it is useful to derive a general relationship that applies to any KT. This relationship describes the degree to which a kernel K with support on the unit interval and kernel transform KT bloats an estimator $\hat{f}$, as measured by the bias

$$\overline{\sigma}^2 = \mathrm{E}[\mathrm{Var}_{\hat{f}}(X)] - \mathrm{Var}_f(X),$$

the difference between the expected variance of the estimated density and the variance of the density that is estimated. This difference provides a particularly good example of one of the peculiarities that characterizes what Tukey (1974) referred to as the elegance of statistical applications of Fourier series-based generalized representation.

In Chapter 3 it was mentioned that the Dirichlet kernel plays a central role with respect to the moments of an estimated density. We will now derive two basic relationships that illustrate this role.

Suppose a curve estimator $\hat{f}$ constructed from n iid variates with datum density f, which has support on the unit interval, is based on the multiplier sequence $\{b_k\}$. The difference between the expectation of the νth moment about the origin of $\hat{f}$ and the corresponding νth moment of f equals

$$\sum_{k\neq 0} B_k(b_k-1)\left\{\frac{1}{2\pi ik}+\sum_{s=1}^{\nu}\frac{\nu(\nu-1)\ldots(\nu-s+1)}{(2\pi ik)^{s+1}}\right\}.$$

The corresponding relationship between the νth moments about any μ is

$$\sum_{k\neq 0} B_k(b_k-1)\left\{\frac{(1-\mu)^\nu-(-\mu)^\nu}{2\pi ik}+\right.$$

$$\sum_{s=1}^{\nu} \frac{\nu(\nu-1)\ldots(\nu-s+1)}{(2\pi ik)^{s+1}}[(1-\mu)^{\nu-s}-(-\mu)^{\nu-s}]\Bigg\}.$$

The $\nu=2$ special case of the above expression, which concerns $\overline{\sigma}^2$, has a particularly simple approximation. Expression (4.2.3), $\hat{\psi}(t) = \int_0^1 \exp[-2\pi it\hat{F}^{-1}(y)]dy$, implies that

$$\hat{B}_k = n^{-1}\sum_{j=1}^{n} \exp(-2\pi ikX_j) = \int_0^1 \exp[-2\pi ik\hat{F}^{-1}(y)]dy,$$

which, in turn, implies that

$$\hat{f}(x) = \int_{X_{(1)}}^{X_{(n)}} \sum_{k=-\infty}^{\infty} b_k \exp[2\pi ik(x-z)]d\hat{F}(z),$$

where $X_{(1)}$ and $X_{(n)}$ are the smallest and largest sample elements, respectively. As is pointed out by expression (10) of Tarter and Raman (1970), this formula is identical to the expression used to define a convolution. This implies that if

$$\mathrm{K}(z) = \sum_{k=-\infty}^{\infty} b_k \exp(2\pi ikz)$$

is a density, then $\hat{f}$ is the density of the sum of two independent variates; where the first variate has density K and the second is a discrete variate with density equal to $1/n$ at each unique sample value, and j/n for any sample value that occurs in the sample $1<j\leq n$ times.

The variance of this second density equals $[(n-1)/n]$ times the sample variance s^2, where s^2 is an unbiased estimator of f's variance. Therefore the expectation of the variance of $\hat{f}$ equals $[(n-1)/n]$ times the variance of the datum density f plus the variance of K. As n increases the term $[(n-1)/n]$ rapidly approaches one. Hence, the bloating $\overline{\sigma}^2$ induced by any kernel is almost exactly proportional to the variance of the kernel K, σ_K^2. Consider a kernel with support on the interval $[-1/2, 1/2]$ such that $\sigma_K^2 = \int_{-1/2}^{1/2} z^2 K(z)dz$ is well defined, K's transform KT is symmetric about zero and $\int_{-1/2}^{1/2} K(z)dz=1$. In this case $b_0=1$, $b_s=b_{-s}$, for $s=1,2,\ldots,$ and

$$\sigma_K^2 = 2\int_0^{1/2} x^2\left[1+2\sum_{k=1}^{\infty} b_k \cos(2\pi kx)dx\right] \tag{5.3.1}$$

$$= \frac{1}{12}+\pi^{-2}\sum_{k=1}^{\infty}(-1)^k b_k k^{-2}.$$

For positive b_k, $k=0,\pm 1,\pm 2,\ldots,$ (5.3.1) is an alternating series. Therefore, the greater the difference between the absolute magnitudes of b_k and b_{k+1} (equivalently, the more steeply KT(z) decreases to zero as $|z|$ increases), the larger σ_K^2 will be. Also, since k^{-2} is much larger for small positive values of k than it is for large values of k, a KT that at first rapidly decreases and then levels off, leaving almost equal consecutive b_k, will tend to have a larger value of σ_K^2 than will a KT that first decreases slowly and then rapidly decreases.

Note that the Dirichlet kernel introduced in Section 3.3 corresponds to a rectangular KT that is symmetric about the point $z=0$. Because of the near identity between σ_K^2 and $\overline{\sigma}^2$, a KT that decreases linearly to zero will inflate the variance of an estimator more than the rectangular KT of a Dirichlet kernel that is horizontal over some neighborhood of $z=0$. Also, a linearly decreasing KT will inflate $\overline{\sigma}^2$ less than some KT like $e^{-|z|}$. A kernel identical to a Cauchy density will have a KT proportional to $e^{-|z|}$. The equally spaced evaluations, b_k, $k=0,\pm 1,\pm 2,\ldots,$, of the KT, $e^{-|z|}$ assume values that differ more for small indices k than they do for large indices. Hence, they induce large values of $\overline{\sigma}^2$. We will see in the next section that the reverse is true; that KTs that increase rather than decrease near the point $z=0$ have applications that are opposite to applications of KT like $e^{-|z|}$. They deflate, rather than increase estimator bloating and $\overline{\sigma}^2$, for this reason, can be used to reduce mixture component overlap.

The role played by multipliers $\{b_k\}$ in the history of mathematics has led most professional mathematicians to study categories of functions not based on specific closed-form expressions. For example, among newly published mathematics books it is only tables of integrals which contain chapters devoted exclusively to functions that are composites of one or two elementary functions like x^k and e^{kx}. Instead, functions are primarily categorized in terms of attributes such as continuity, differentiability, whether they have a well-defined integral over some interval, for example, if the interval is the entire real axis, the curve is classed as an L_1 function, or whether they have a well-defined integral when squared, for example, are L_2, as discussed in Chapter 2. The reasoning that underlies the shift toward a more general approach stems, in large part, from studies of the peculiarities of Fourier expansions.

The function $f_l(x) = I_{[0,\alpha]}(x)[x^{-1/2}]/[2\alpha^{1/2}]$ illustrates one such peculiarity. While $f_l(x)$ is a density (and therefore can be integrated over any interval) $[f_l(x)]^2$ *cannot* be integrated over any region that contains or is bordered by the point zero. This shows that $f_l(x)$ is an L_1, but not an L_2, function.

L. Carleson showed in 1966 that over all sets of points (except possibly those of measure zero) any truncated $2m+1$ term Fourier expansion f_m of an L_2 function f converges to f as $m \to \infty$. But, as $f_l(x)$ demonstrates, there can exist densities that are not L_2. Therefore, with respect to Carleson's criterion, a slight gap exists in the universality or generality of the density estimation applications based on Fourier analysis.'

The existence of gaps such as that characterized by $f_l(x)$ was studied by the mathematician whose name is borne by the most important of the many kernels and multiplier sequences used in the mathematical literature. What are known as the Dirichlet conditions were derived by the mid-nineteenth century mathematician Peter Gustav Lejeune Dirichlet. It was Dirichlet who established the first rigorous mathematical underpinnings for the representational approaches invented by Fourier in the early-nineteenth century.

If a function f is: (1) uniquely defined on [0,1], and (2) bounded on [0,1], (3) such that any two consecutive points on [0,1] at which f is discontinuous are separated by an interval of length $C>0$; and (4) where for any such interval it has only a finite number of maxima and minima; then f satisfies the Dirichlet conditions. Dirichlet showed that these conditions ensure that the Fourier expansion of f uniquely converges to f at all but some set of points of measure zero.

But, suppose a function satisfies neither the Carleson nor the Dirichlet conditions? In this case Fejér's and other related expansion techniques can be used to extend the applicability of expansion methods. These techniques play such a prominent role in the mathematical literature that the entire book *Summable Series and Convergence Factors* (Moore 1938) deals with this single topic.

Fejér's approach is as follows: Suppose f is a function defined on the unit interval. Define a function $\bar{f}_m$ to equal

$$(m+1)^{-1}\sum_{u=-m}^{m} f_u,$$

where for the nonnegative integer u, the expression known as the uth Fejér partial sum,

$$f_u = \int_0^1 f(t)\mathrm{K}_{\mathrm{D}(u)}(x-t)dt,$$

is the integral over the unit interval of the product of the function to be expanded times the u-term Dirichlet kernel $\mathrm{K}_{\mathrm{D}(u)}$ evaluated at $x-t$. Fejér's $\bar{f}_m$ is the fraction whose numerator is the total of the zeroth through mth Fejér partial sum. Note that the first term of each f_u's expansion, $[B_1\exp(2\pi ikx)]$, is repeated in all m partial sums.

Dividing the numerator of $\bar{f}_m$ by $m+1$, and then gathering together the m terms $B_1\exp(2\pi ix)$, yields an $\bar{f}_m$, that equals 1, plus the quantity $[1-\{1/(m+1)\}]B_1\exp(2\pi ix)+\ldots$. When repeated, this process yields the truncated series

$$\bar{f}_m(x) = \sum_{k=-m}^{m} [1-\{|k|/(m+1)\}]B_k\exp(2\pi ikx). \qquad (5.3.2)$$

Each square bracketed term of (5.3.2) can be treated as a member of the multiplier sequence $\{b_k = [1-|k|/(m+1)]$, if $-m\leq k\leq m$, and $b_k=0$ otherwise$\}$.

The truncation point m plays two roles, which can be interpreted as follows: In reference to the triangularly or conically crowned hat-shaped KT whose equally spaced evaluations yield $\{b_k = [1-|k|/(m+1)]$, m determines the rate at which the KT hat's crown declines to zero. Correspondingly, $2m+1$ determines just where the KT crown meets the KT brim. Fejér showed that as m increases, expansion (5.3.2) will approach, and in this sense *represent*, any L_1 function and hence any pdf.

Despite the historical importance of the Fejér multipliers, they are of far less value in exploratory curve estimation than are the set of multipliers designated by the Greek letter lambda. For exploratory curve estimation Fejér multipliers can be considered both an aid to the understanding of the general role of multipliers in data analysis and as an historically important step leading to the lambda multiplier system.

Before turning to topics concerning the lambda KT, it will be useful to consider the application of Dirichlet's conditions to a statistical density. The failure of a statistical density to meet the Dirichlet conditions, although possible, is not likely in practice. The two most troublesome Dirichlet conditions are (3) and (4). Within some small interval of length C, these conditions describe the irregularity of a density. For any continuous variate all actual data used as digital computer input must be of limited precision. Hence, the smallest possible interval, LMIN, within which it is possible to distinguish two distinct data points will always be some finite length. While it is theoretically possible that two consecutive points at which a density f is discontinuous will *not* be separated by an interval of length $C > 0$, if LMIN is some nonzero real number there is no way that any sample of data can be used to establish directly whether or not Condition (3) is met. No data set can be gathered with enough precision to allow the validity of this condition to be determined empirically.

This suggests that even though the Fejér multiplier sequence applies to the hypothetical case where sample size n of infinitely precise data points approaches infinity this sequence will not necessarily be useful when problems involve finite samples of data gathered to a finite degree of precision. In fact, we have often found empirically that for those few applications where a Fejér sequence is of some value, some special case of the lambda multiplier sequence is equally, or even more, useful.

As previously mentioned, the Fejér multiplier sequence can be interpreted in terms of the evaluations of a conical hat-shaped KT whose outline is an isosceles triangle between $-m$ and m, the hat's crown, and is equal to zero elsewhere, the hat's brim. We will see below that the Dirichlet KT is the special case of the lambda KT for $\lambda = 0$ that corresponds to the Wahba raw estimator introduced in Section 3.2. While the region beneath a two dimensional Dirichlet KT has the silhouette of a top-hat with a brim of infinite length, the Fejér silhouette resembles that of a conical witch's cap, again with a brim of infinite length.

The kernel $K_F(z)$ whose Fourier transform, KT, has a triangular shape is

$$\sin^2(mz/2)/[m\sin^2(z/2)]. \tag{5.3.3}$$

Since m is positive and the sin function which appears both in the numerator and the denominator of (5.3.3), is squared, $K_F(z)$ must be an everywhere nonnegative function. Therefore, variance inflation bias will be positive. Consequently, applications based on the kernel $K_F(z)$ will fatten, bloat or smooth a density estimator. The conical part of the witch's hat, $1-|k|/(m+1)$ will have the sharpest cone, and thus affect (5.3.1)'s speed of convergence most when m is small. Thus, it is small values of m that most bloat an underlying curve, while as m increases, the angle between the two sides of the isosceles triangle will approach 180 degrees. Figuratively speaking, the cone of the witch's hat will then approach the flat top of the Dirichlet kernel transform, $KT_D(z)$, top-hat and have minimal variance inflating bias $\overline{\sigma}^2$.

5.4 Lambda multipliers

We now have completed the background needed to describe a sequence of equally spaced KT evaluations that can greatly expedite the mixture decomposition process. Suppose $K_F(x)$'s transform was modified to form KT_λ, which rather than being triangular or conical has the shape of a truncated normal density. The Dirichlet and Fejér kernels, $K_D(x)$ and $K_F(x)$, have KTs whose form is very different from the forms assumed by $KT_D(z)$ and $KT_F(z)$. However, the characteristic function or Fourier transform of a normal distribution symmetric about zero is itself proportional to a normal density. Nevertheless, owing to normal curve truncation, there is a subtle distinction between KT_λ and its Fourier transform K_λ.

It is theoretically possible to modify the triangular Fejér KT for use in mixture decomposition applications. However, to our knowledge the approach described below has so far only been studied in the normal-reciprocal normal case, which we will call, “lambda methodology.” Nevertheless, to illustrate how lambda works, it will be useful to begin with a detailed discussion of the Fejér kernel.

Consider that the Fejér KT can be treated as the product of two functions, the first defined for positive arguments as a slanted line of negative slope (which borders the triangular or conical part of the witch's cap), and the second the indicator function $I_{[-m,m]}(z)$, which ensures that the brim of the hat will consist of horizontal rays of zero height within the complement of the interval $[-m,m]$. Thus, the Fejér KT can be treated as the product of $I_{[-m,m]}(z)$ times a continuous function that for positive arguments is a straight line with slope $-1/(m+1)$ and for negative arguments is a straight line with slope $1/(m+1)$, where both lines have a y-intercept equal to 1.

Tolstov (1962, page 124) presents the useful formulas

$$\alpha_u = (a_0A_u)/2+(1/2)[a_v(A_{u+v}+A_{v-u})+b_v(U_{u+v}+U_{v-u})], \quad (5.4.1)$$
$$\beta_u = (a_0U_u)/2+(1/2)[a_v(U_{u+v}-U_{v-u})-b_v(A_{u+v}-A_{v-u})], \quad (5.4.2)$$

for the uth Fourier coefficient $\alpha_u+i\beta_u$ of the product of two functions g and h, where $\{A_k+iU_k\}$ and $\{a_k+ib_k\}$ are the kth Fourier coefficients of g and h, respectively. For example, the Fejér KT is a product of two functions, one of which is a line for positive arguments and one of which is the indicator function. Hence, the Fourier coefficients of the Fejér KT, can be conceptualized as the result of applying (5.4.1) and (5.4.2) to compute each Fourier coefficient $\alpha_u+i\beta_u$ in terms of the Fourier coefficients of these two functions.

Similarly, the kernel that corresponds to the lambda approach to mixture decomposition can be graphed by means of (5.4.1) and (5.4.2). Like the Fejér KT, the symmetric truncation about the point zero of any function $h(x)$ beyond the points $z=\pm m$ is equivalent to the multiplication of $h(x)$ by the indicator function $I_{[-m,m]}(z)$. The Fourier coefficients of the lambda KT's Fourier expansion are the equally spaced evaluations of the lambda kernel. These can be determined by using (5.4.1) and (5.4.2) to calculate the Fourier coefficients of the product of two functions; the first of which has as its Fourier coefficients the equally spaced evaluations of the reciprocal of a normal density symmetric about zero, and the second of which is an indicator function also symmetric about zero.

Expressions (5.4.1) and (5.4.2) can be used to calculate the Fourier coefficients of many types of truncated curve, besides the lambda KT. For example, suppose the untruncated form $f_\lambda(z)$ of some mixture results from the execution of the lambda decomposition procedure. It is shown in Tarter (1979b) that to graph a truncated modification of the function

$f_\lambda(z)$ that zeros out all but one of the components of this mixture over the complement of some interval (c_1,c_2), $\{A_k+iU_k\}$ of (5.4.1) and (5.4.2) can be set equal to $f_\lambda(z)$'s Fourier coefficients, and $\{a_k+ib_k\}$ can be set equal to the Fourier coefficients of the indicator function $I_{[c_1,c_2]}(z)$. Of course, if the resulting curve is to be used as a density estimator, it is also necessary to determine a constant, which when multiplied by the truncated form of $f_\lambda(z)$ ensures that the estimated density integrates to one over the z-axis.

In summary, the following are two distinct ways that (5.4.1) and (5.4.2) can be used to increase the applicability of lambda-based curve contrast control:

(1) As shown in Chapter 3, there is a close correspondence between kernel-based density estimation and the form of series-based exploratory curve estimation emphasized in this book. Consequently, kernels that, with certain settings, can either reduce (as do all nonnegative kernels) but also *enhance* contrast, can be determined by making use of the fact that the Fourier coefficients of the lambda KT's Fourier expansion are the equally spaced evaluations of the lambda kernel. The evaluations of these lambda kernels can even be multiplied by the evaluations of some of the commonly used kernels graphed in Chapter 3 to yield new kernel variants.

Note that in this application of (5.4.1) and (5.4.2), use of these two formulas is equivalent to evaluating the truncated Fourier series with lambda multiplier coefficients over an equally spaced sequence of points. Also, if the Fourier coefficients of some commonly used kernel K are multiplied by lambda multipliers, the resulting products can be used to form a composite series estimator. Alternatively, the Fourier series that has these product multipliers as its coefficients can be graphed in order to examine the effect produced on the original kernel by combining it with the contrast modification induced by the lambda K.

(2) As will be detailed in the next sections, when used for mixture decomposition applications, (5.4.1) and (5.4.2) can be used to tease apart individual mixture components.

Consider the following property of the Fourier transform: If some density $f(x)$ is a mixture of the c densities f_s, $s=1,\ldots,c$

$$f(x) = \sum_{s=1}^{c} p_s f_s(x), \qquad (5.4.3)$$

then the Fourier transform (or characteristic function) of f will be a mixture of the Fourier transforms, characteristic functions, of the c mixture components f_s. Also, the kth Fourier coefficient of the expansion of f will be the weighted average, with weights p_s, of the c,kth Fourier coefficients of the c components f_s, $s=1,\ldots,c$.

Assume that for $s=1,..,c$ the sth characteristic function of $f_s(x)$ can be expressed as $\exp[W_s(it)]$ in terms of the cumulant generating function $W_s(z)$, where the *jth* cumulant κ_{js} of the *sth* component is assumed to exist and be well defined. (The *sth* cumulant is the evaluation at zero of the *sth* derivative of the log of the MGF.) Then by the definition of cumulants given by Kendall (1952, page 90)

$$W_s(z) = \sum_{s=1}^{c} \kappa_{js} z^j / j!. \tag{5.4.4}$$

For example, if $f_s(x) = \phi[(x-\mu_s)/\sigma_s]/\sigma_s$, $\phi(x) = N(x|0,1)$, then

$$W_s(z) = \mu_s z + \sigma_s z^2/2. \tag{5.4.5}$$

Combining the above relationships, suppose a certain sample of iid data with datum density f has been transformed to the unit interval in such a way that for all practical purposes the *kth* Fourier coefficient of the expansion of f equals

$$B_k = \sum_{s=1}^{c} p_s \exp\left[-2\pi i k \mu_s - 2\pi^2 k^2 \sigma_s^2 - \sum_{j=3}^{\infty} \kappa_{js} (ik)^j / j!\right]. \tag{5.4.6}$$

Now, for the sequence of multipliers

$$\left\{b_k^{(\lambda)} = \exp\left[2\pi^2 k^2 \lambda^2\right]\right\}, \tag{5.4.7}$$

when f_λ is defined to be the function whose Fourier expansion has coefficients $\{b_k^{(\lambda)} B_k\}$, it follows that each member of this sequence of coefficients equals

$$\sum_{s=1}^{c} p_s \exp\left[-2\pi ik\mu_s - 2\pi^2k^2(\sigma_s^2-\lambda^2) - \sum_{j=3}^{\infty} \kappa_{js}(ik)^j/j!\right]. \quad (5.4.8)$$

To interpret the effect of the multiplier sequence (5.4.7) on the function whose kth Fourier coefficient is B_k, notice that the one difference between expressions (5.4.6) and (5.4.8) is that in (5.4.8) the term $(\sigma_s^2 - \lambda^2)$ appears where the term σ_s^2 had appeared in (5.4.6).

The roles played by lambda and other multiplier sequences described below can be explained by noting that the term $(\sigma_s^2 - \lambda^2)$ of (5.4.8) is associated with an important distributional property of the underlying mixture model. With one exception $(\sigma_s^2 - \lambda^2)$ is the second cumulant, variance, of the sth component of a mixture identical to f. This exception is that the original second cumulants of each original component have each been reduced by the quantity λ^2. Below we will refer to the application of this connection between the second cumulant and $(\sigma_s^2 - \lambda^2)$, as "contrast enhancement."

Consider the circumstances in which it would be easy to decompose a mixture. If the second cumulants of all distributional components were very small in comparison to the spacings between the component first cumulants, for example, μ_1 and μ_2 of $f(x|p,\mu_1,\mu_2,\sigma)$ for standard normal f_1 and f_2, then with any moderate sized sample of data, any of the graphical methods described in earlier parts of this chapter could be used to easily separate components.

Conversely, suppose there were only two components, but that the second cumulants of these components are large in comparison to the space between the two means. Then a very large sample of data and a highly effective mixture decomposition procedure would be needed to obtain useful information concerning the structure of the mixture. Thus, the lambda method deals with a mixture in a manner analogous to the injection or oral application of many prescription drugs. It targets precisely those properties of all components, the second cumulants, whose reduction greatly facilitates the solution of the mixture decomposition problem.

5.5 The lambda KT used for $\overline{\sigma}^2$ reduction

The sequence of multipliers $\{b_k^{(\lambda)} = \exp[2\pi^2k^2\lambda^2]\}$ differs substantially from Fejér and Dirichlet multiplier sequences, in the sense that $\{b_k^{(\lambda)} =$

$\exp[2\pi^2k^2\lambda^2]\}$ *increases* rather than decreases to the left and to the right of the index $k=0$, where it takes on the value one, as do the Fejér and Dirichlet sequences. That this sequence increases as k increases complicates the process of expressing the lambda KT's Fourier transform. For positive real valued λ, $b_k^{(\lambda)} \to \infty$ as $k \to \infty$. However, the index m of the last term used by a truncated series estimator is finite. Thus, like the KT that corresponds to the Dirichlet and Fejér kernels, in practice lambda multipliers are equally spaced evaluations of a function which has a finite support region. Within this region the lambda KT is formed by removing the minus sign from the argument of the normal curve exponent $-(z/\lambda)^2/2$. The lambda KT equals $\exp[(z/\lambda)^2/2]$ within the interval $[-m,m]$. This function increases rather than decreases as $|z|$ increases and then, outside of $[-m,m]$, is identically equal to zero. In terms of hat shape, the crown of a lambda hat is parted in the middle like a Fedora which has a brim of infinite length. Examples of the kernels formed from these sequences are shown in Figure 5.7.

To clarify the processes by which the lambda KT can be applied, consider the following example: Suppose one wishes to estimate the $c-1$ unknown mixing and c unknown location parameters of a mixture where all components are Cauchy densities with a common *known* scale parameter σ. (The standard Cauchy g with $\sigma=1$ is defined to equal the Student-t density with one degree of freedom, $g(x)=\{\pi[1+x^2]\}^{-1}$.) Although the Cauchy does not have well-defined cumulants, it *does* have scale parameter that can be defined in precisely the same way the scale parameter of a normal can be defined. Specifically, if $f(z)$ were to represent the standard form of either the normal or Cauchy, then $f([x-\mu]/\sigma)/\sigma$ would represent the general form with location parameter and scale parameter where neither parameter of the general Cauchy has any connection with the second cumulant. For any Cauchy, the odd cumulants beginning with $E(x)$ are undefined in the sense that different values can be attained, depending on the definition of the definite integrals used to define the odd cumulants. In addition, all even cumulants are infinite. Thus, for the Cauchy case, there can be no connection between the scale parameter and the second cumulant.

Even though the connection between scale parameters and second cumulants is one of the sufficient conditions for lambda method applicability, it is not a necessary condition. Suppose, for example, a mixture of Cauchy densities has a common and known scale parameter σ of each component density. Let $\{b_k^{(C)}\}$ represent a sequence of equally spaced evaluations of the reciprocal of characteristic function of the

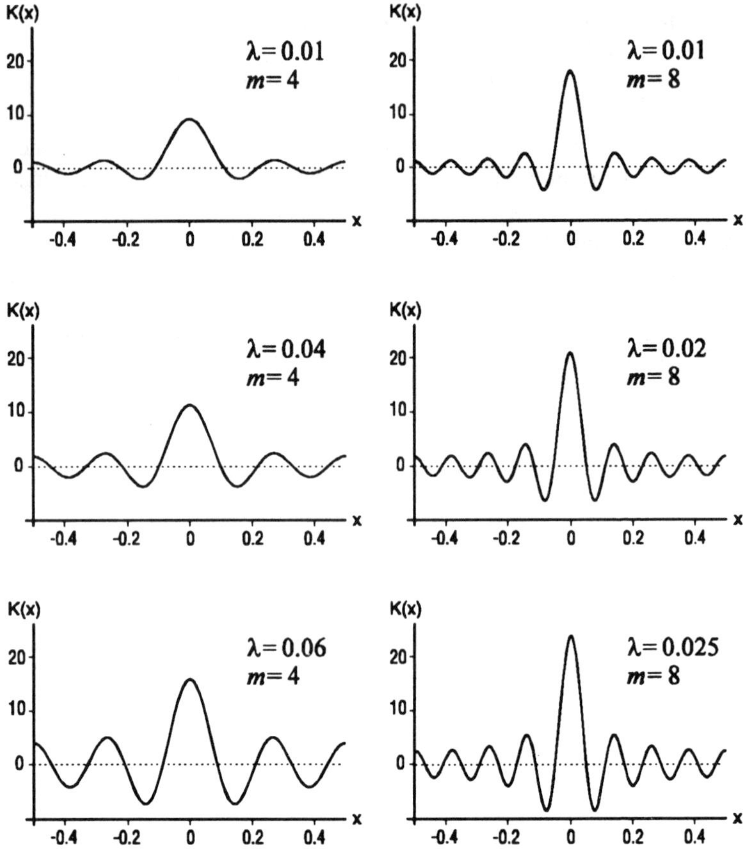

Figure 5.7. *Examples of λ kernels for three choices of the user setting λ and two truncation points m.*

truncated Cauchy within some finite interval that contains zero. Each term of $\{b_k^{(C)}\}$ can be multiplied by the Fourier coefficients of a normal with scale parameter σ, and these products can be used to construct a series curve estimator. Specifically, by using the multiplier sequence $\{b_k^{(CN)} = \exp[2\pi(|k|\sigma-(k\sigma)^2/2)]\}$, any mixture of c Cauchy densities can be transformed to a mixture of c normal densities. The factorization of the assumed common parts of the Cauchy mixture components from the expression for the characteristic function of the Cauchy mixture ensures that by using the multiplier sequence $\{b_k^{(CN)}\}$ Cauchy density components can be replaced by normal components *within* the mixture *without* first decomposing the mixture.

When evaluated over the positive real axis, the Cauchy cumulant generating function (CGF) is the linear function of z, $(i\mu+\sigma)z$. Thus, because this CGF contains no quadratic and higher-order terms, use of a variant of the lambda method to decompose a Cauchy mixture requires that one factor be the constant e raised to some linear function. This, in turn, requires that not only must all mixture components have a common scale parameter, but also that the value of this parameter be known. We will now consider another lambda variant that involves prior knowledge of distributional structure.

Suppose a mixture is comprised of components where all order-three and higher-order cumulants κ_{js}, $j=3,4,5,\ldots$, $s=1,2,\ldots,c$ exist. Suppose further that (1) for $s=1,2,\ldots,c$, the cumulants have common known values $\{\kappa_{js}=\kappa_s,\ j=3,\ldots\}$, but (2) that for each mixture component the values of the first and second cumulants are different, but unknown. Assumptions (1) and (2) imply that

$$B_k = \sum_{s=1}^{c} p_s \exp\left[-2\pi ik\mu_s - 2\pi^2k^2\sigma_s^2 - \sum_{j=3}^{\infty} \kappa_{js}(ik)^j/j!\right]$$

$$= \left[\exp\left[-\sum_{j=3}^{\infty} \kappa_j(ik)^j/j!\right]\right]\left[\sum_{s=1}^{c} p_s \exp\left[-2\pi ik\mu_s - 2\pi^2k^2\sigma_s^2\right]\right].$$

Therefore, by means of the multiplier sequence

$$\left\{b_k^{(\kappa)} = \exp\left[\sum_{j=3}^{\infty} \kappa_j(ik)^j/j!\right]\right\} \tag{5.5.1}$$

the problem of decomposing *any* mixture that has the above structure can be reduced to the problem of decomposing a mixture of c normals, where the values of c, the mixing parameters $\{p_s,\ s=1,\ldots,c\}$, and the $2c$ normal location and scale parameters are all unknown.

The above methods illustrate how the factorization that underlies the lambda technique can be applied even when component densities are not symmetric functions. However, although many problems can be solved that are more difficult than the decomposition of $c=2$ normals, second cumulant reduction by the lambda method is a much more useful data analysis tool than is methodology based on $\{b_k^{(\kappa)}\}$ of (5.5.1). This is because even assuming the unlikely event that a mixture has components that share a great deal of common structure, it stretches credulity that the exact nature of this structure will be known. Thus, although the lambda procedure can be generalized to deal with mixtures comprised of nonnormal and even asymmetric densities with nonzero odd higher order cumulants, the most useful application of this procedure is to the decomposition of mixtures of components that, like the normal, are symmetric.

5.6 Generalizations and limitations of lambda methodology

While it has many uses and generalizations, lambda methodology also has certain peculiarities and limitations. Among its practical applications is the capacity to create scatter plots and stick pin maps, that enhance contrast. We will begin this section with a brief discussion of this application, followed by a review of the comparative practical advantages and disadvantages of mixture decomposition approaches.

There are two requirements for the creation of scatter plots and stick pin maps that enhance contrast. (1) The mathematical representation of a contrast reducing kernel must be available in a computationally tractable form. (2) The contributions of the KT equally spaced evaluations (5.4.7) that provide the Fourier coefficients of a contrast enhancing kernel must be used to tally both the positive and the negative contributions made by a single data point to the overall display.

To better interpret requirement (2), consider that any plotting character that spreads a single intensity plotting character over a two dimensional space is the counterpart of a rectangular kernel. The KT that is rectangular does have as its transform the neutral variance inflating bias $\overline{\sigma}^2$ Dirichlet kernel. Yet the kernel K that corresponds to a rectangular function is very different from its KT. Specifically, a rectangular kernel corresponds to a KT that is shaped like the Dirichlet kernel. However, it is the Dirichlet kernel and not a Dirichlet kernel shaped KT that is connected with neutral variance inflating bias $\overline{\sigma}^2$. Unlike the Dirichlet kernel, any solid plotting character will always inflate $\overline{\sigma}^2$, and often inflate $\overline{\sigma}^2$ substantially.

The inflation of $\bar{\sigma}^2$ has the effect of smoothing over or increasing the overlap between distributional subcomponents. However, what is required in many applications is not smoothing or contrast reduction, but instead, contrast enhancement involving the reduction of subcomponent overlap. The question then is: how can the multiplier techniques described in the previous sections be used to devise a physical or electronic process for contrast enhancement? In actual practical applications all that is needed is an actual, or electronic, analog of a fine spray consisting of small black dots, for positive kernel values, and white dots, for negative kernel values, on a gray background, where dot-spray density is made proportional to lambda kernel height. In this fashion both positive and negative kernel values can be displayed.

A detailed description of the dot-spray technique is given by Tarter and Freeman (1988). There are obviously many feasible dot-spray variants based on color choice as well as the selection of plotting character shapes and sizes to convey the impression of negative as well as positive kernel evaluations. In particular, we have found that blanking out alternate terraces of total spray contributions can provide a very computationally efficient alternative to conventional contouring programs. Considered from a general perspective, the physical realization of a kernel can be a specified array of icons, where each icon is associated with some negative or positive real number. This array serves to generalize the conceptualization, data *point*. The word point is actually a misnomer for a rectangular kernel anyway, so why not treat any data "point's" number pair, or any such pair's higher dimensional generalization, as merely some kernel's location? Since many modern printing and electronic display techniques are digital in nature, it is as reasonable to use a spray of identical icons as it is to use a uniformly spread, single icon, to represent a datum's contribution to the overall display. This last assertion has particular relevance to graphical displays that are based on small or moderate sized samples.

Through implementation by the lambda-based spray technique, it is easy to see how a spray of black and white dots, or many other icon systems, can be used to enhance, rather than reduce, contrast. Although the resulting process raises issues related to optimal kernel design, it differs in an important way from most currently used curve estimation procedures. This is because the goal of contrast enhancement is not the same goal aimed at by estimates of the underlying density of data elements. Instead, the goal of contrast enhancement is to display some hypothetical density that contains mixture components which all have reduced second cumulants. From a theoretical, rather than a practical, perspective, within the metric used to evaluate the statistical process, multipliers modify *both* the estimator, and the curve that is estimated.

It is informative to examine the actual structure of the metric that underlies the lambda contrast enhancement process

$$J(\hat{f},f) = \mathrm{E}\int_0^1 \left[\sum_{k=-\infty}^{\infty} b_k^{(\lambda)} b_k^{(J)} \hat{B}_k \exp(2\pi ikx) - \sum_{k=-\infty}^{\infty} b_k^{(\lambda)} B_k \exp(2\pi ikx) \right]^2 dx,$$

where $\{b_k^{(J)}\}$ is determined using the optimization methods described in the preceding chapter. What is noteworthy about this metric is that the density f that is estimated is not the datum density. Instead, it is some artificial density, which differs from the datum density in the sense that the contrast between mixture components has been enhanced. This departure from the usual metric that does contain the datum density as one of its arguments corresponds to the presence of the multiplier $b_k^{(\lambda)}$ within the representation of the estimated curve and the composite multiplier $b_k^{(\lambda)} b_k^{(J)}$ within the representation of the curve estimator. We will show in Chapter 8 that for the above unweighted MISE metric, the chice of $b_k^{(J)}$ can be made without regard for the choice of the setting which governs the value assumed by each multiplier $b_k^{(\lambda)}$. This invariance of curve choice and curve estimator choice greatly expedites the computational processes that underlie Fourier series-based data analysis.

One aspect of contrast enhancement in bivariate and higherdimensional spaces does not apply to the univariate case. Advantage can be taken of the geometric position of components. Tarter and Silvers (1975) describe the details of a bivariate lambda procedure by which the twodimensional geometry of the mixture model can be used to tease apart components by selecting particular choices of the variance-covariance matrix generalization of univariate λ^2. For example, in situations where the major axes of two or more components are parallel, certain choices of the variance covariance matrix can be used to reduce the length of the minor axes of two superimposed components. In a manner analogous to the careful cleavage of a raw diamond, the right choice of cleavage angle allows multivariate mixtures to be separated with smaller degrees of contrast enhancement (analogous to lighter blows of the diamond cutter's hammer) than would generally be needed in the univariate case.

This is because in the univariate case, components can only be drawn apart in one dimension. Although from a theoretical point of view it is possible to decompose mixtures more effectively in the bivariate, than in the univariate, case, there may nevertheless be a great many of computational difficulties encountered when data sets containing high-dimensional data points are studied.

Besides generalizing lambda methodology to higher-dimensional spaces, it is possible to generalize the contrast-enhancement process by simultaneously determining the KT truncation point m and the contrast modification setting λ. For any subinterval of the unit interval, (5.3.1) and (5.3.2) can be used to determine the Fourier coefficients of any kernel or KT truncated within the interval $[-m,m]$ in terms of the Fourier coefficients of the untruncated kernel or KT. Alternatively, explicit formulas for the evaluation of characteristic functions of the truncated normal, Cauchy, Laplace, logistic and gamma densities are given by Kronmal and Tarter (1968, expressions (62) through (68)).

Similarly, for some $0<m_2<m_1$ it is useful to generalize the Fejér multiplier sequence $[1-|k|/(m+1)]$, $k=0,\pm1,\pm2,\ldots,\pm m$ to $[1-|k|/(m_1+1)]$, $k=0,\pm1,\pm2,\ldots,\pm m_2$. This generalization can be interpreted in terms of the equally spaced evaluations of a symmetric, but truncated, triangular KT whose top resembles the roof and whose edges resemble the sides of a house. Through truncation the tendency of the Fejér kernel to bloat the curve estimator can be controlled in much the same way that selection of specific values of λ can be used to control bloating. In addition, the simultaneous choice of m_1 and m_2 allows the Fejér KT to be used in conjunction with the formulas described at the beginning of Section 5.3 to control estimated second moment bias. This is because, as the modulus of Fourier coefficient B_k decreases, the variance of its estimator increases. Thus, to estimate second moment bias accurately, it is best to use a short sequence of series terms, in other words, a small value of m_2. Once this small value is selected, the remaining user control setting m_1 can be used to modify second moment bias.

When the weight w of the MISE metric is uniform, the composite curve estimator upon which the lambda method is based has the following useful property: No matter what value of λ is substituted in $\{b_k^{(\lambda)} = \exp[2\pi^2k^2\lambda^2]\}$, the sequence of estimated multipliers need not be altered to minimize the MISE which assesses the fit of the contrast-enhanced estimator to the curve that it estimates. Thus, by using lambda series formulation, the contrast of an estimator can be varied by means of easily implemented modifications of the same techniques previously applied to the problem of conventional series-based density estimation.

Although there are many uses to which lambda methodology can be put, it must be cautioned that, like most other general statistical techniques, lambda multipliers have their own peculiarities and limitations. The most important limitation can be explained in terms of the cumulant-based expression

$$\sum_{s=1}^{c} p_s \exp\left[-2\pi i k\mu_s - 2\pi^2 k^2(\sigma_s^2 - \lambda^2) - \sum_{j=3}^{\infty} \kappa_{js}(ik)^j/j!\right]. \quad (5.6.1)$$

It is obvious that as $n \to \infty$, the resulting estimator will degenerate unless $(\sigma_s^2 - \lambda^2)$ is positive for all indices s from one to c. In addition, unless for all $1 \le s \le c$ all κ_{js} are zero for $j > 2$, it may be the case that even though λ^2 is less than all distributional second cumulants, some choice of λ^2 will generate a sequence of cumulants of a function which is not everywhere nonnegative.

Even if the lambda method produces contrast enhanced estimators that take on negative values, (5.4.1) and (5.4.2) can be still be used to decompose a mixture. In many applications it is feasible to use the lambda method after reversing the sign of λ to cancel out the effect of the original contrast-enhancement process *after* eliminating overlap with other distribution components. By using this procedure separate optimal multiplier sequences can be estimated for each separated sequence of component-density estimated Fourier coefficients. Finally, a series estimator of the mixture density can be obtained by using subsequences of coefficients based on separately optimized estimated multipliers. Details of this process are described in Tarter (1979b).

Because it is necessary for all terms $(\sigma_s^2 - \lambda^2)$ within the expression (5.6.1) to be positive, it is best to use the *smallest* value of λ that lowers the degree of mixture component overlap to insignificance. For most sample sizes, it is not advisable to degenerate mixture components completely, by using a λ value close to the value of the smallest σ_s, $s=1,\ldots,c$. Instead the following four step procedure should be used:

(1) Find the smallest λ which causes the apparent component overlap to vanish at a single point. In other words, this λ is the largest value such that the support of the second cumulant reduced component can be taken to be wholly within some subinterval M. Note,

this step involves the interplay of two effects that sample size, n, has on the lambda process.

Series estimated from samples where n is small will commonly be stiff, in the sense of being low-degree trigonometric polynomials, rather than longer expansions containing terms with many oscillations within a given interval. As a result of this stiffness we have found that the values of λ that zero out apparent overlap tend to be small. On the other hand, as n increases, the best choice of λ will approach a limit equal to the smallest of the mixture component standard deviations and the estimated contrast enhanced curve will approach a density that is zero outside the neighborhood of the degenerate interval M consisting of a single point. However, in actual practice we have found it best to avoid choices of λ that cause the contrast enhanced curve to take on negative values under all but the extreme tail support subregions.

(2) Within the complement of M damp-out the effect of extraneous mixture components by using (5.4.1) and (5.4.2). If the components to be excised are located in the tail region, it is best to excise over the entire tail region even when the estimated curve takes on negative values. The damping or excision process can be accomplished by using (5.4.1) and (5.4.2) to determine the Fourier coefficients of the product of the displayed λ-reduced estimator times the function $I_M(x)$, which equals zero over the complement of M.

(3) Reverse the λ-reduction process by using multipliers respectively equal to the reciprocals of the corresponding members of the $\{b_k^{(\lambda)} = \exp[2\pi^2k^2\lambda^2]\}$ sequence,

(4) Continue the preceding three steps as needed to further separate distributional components.

Henna (1985) and Silverman (1981) describe a mixture decomposition hypothesis test that can be used to determine the number of components, c, within mixtures that are *known* to be composed of normals. Like these approaches, the lambda method *does* assume a mixture model of the population density. However, instead of assuming a parametric structure for the mixture, it assumes that the first and second cumulants take on values that make it possible for second cumulant reduction by λ^2 to reduce subcomponent overlap.

Unlike Henna's approach, effective use of the lambda method does not require that components be exactly normal. Note that, as illustrated in the previous section, for purposes of contrast enhancement, normality is equivalent to the assumption that all third- and higher-order cumulants are zero. Thus, normality is merely a convenience which guarantees that the limiting case of the contrast enhanced curve has the same elementary function representation as does all intermediate cases. But the lambda method is inherently nonparametric, in the sense that it is a *property*, the second cumulant, of a curve, and not any parameter of some representation of that curve, which is altered.

Besides the lambda method's nonparametric formulation, this fundamental process differs from Henna's parametric, albeit useful, approach in a second respect. While the Henna procedure is designed to be sensitive to the value of c, in order to test hypotheses concerning this value, the lambda approach can be implemented without prior information or hypotheses concerning c. Hence, it provides clues, but not necessarily evidence, concerning the value of c. Nevertheless, in conjunction with methodology described in the next chapter, it can be used to check the validity of the assumption of component normality. For this reason lambda methodology can serve as a useful adjunct to Henna's procedure. Also, although at present there is no inferential literature based on lambda multipliers, it will be obvious when we turn to the question of inference in Chapter 10 that there is no theoretical reason why inference cannot be based on these multipliers.

Although, in theory, lambda can be implemented without regard to the numerical value of c, in practical applications the number of mixture components does play an important, although indirect, role. The more components there are, the greater the chance that contrast enhancement will reduce some second cumulant to a value less than zero. Hence, it is best to take advantage of the different degrees of separation between subcomponents. Generally speaking the sample trigonometric moments $\{\hat{B}_k\}$ computed from the sample of data should be allocated among distinct subsequences in a manner which at each step requires the smallest possible value of λ for the cleavage of $\{\hat{B}_k\}$.

Note, the methods described above are quite different from procedures based on density multimodality. An excellent review of the latter is given by Izenman and Sommer (1988).

CHAPTER 6

Curve estimation approaches to model and transformation selection

This and the next chapter will address the following two questions: (1) How can the most appropriate model be selected? (2) How should data be modified by a log, square root or other transformation to a form which is suitable for analysis by elementary function representation? To provide answers to these questions, nonparametric estimates of the two basic curves—pdf, f, and cdf, F—will be used to make decisions concerning two other curves, some standard density model g and some preliminary data transformation v. We will continue to distinguish the word transformation, as in log data transformation, from the word transform, as in Fourier transform. In keeping with the approach emphasized in this book, on the one hand, a transformation, such as the log or square root, and, on the other hand, a pdf model g, such as the standard form normal or extreme value density, will be treated as two elementary function components of a single expression. We will do this even though the log function is often used to modify data points, and the normal is used for a quite different purpose, as a representation of the density of the modified points.

Section 6.1 will explain why treating g and v as merely two components of a single expression underscores a departure from conventional transformation selection procedures that is made possible by new curve estimation approaches. Conventional selection techniques almost always depend on trials of particular transformations such as $Y' = \log(Y-C)$ or $Y' = \sqrt{Y+C}$, where some curve estimator like the cumulative polygon is constructed from samples comprised of values of the post-transformed variate Y'. Use of graphical procedures to a make decisions about the choice of v as a log or alternatively a square root elementary function, without prior knowledge or even trial values of C illustrates a core idea behind many graphical methods. This idea was illustrated by the

Weichelberger (1961) and Bhattacharya (1967) mixture decomposition technique described in the last chapter. The partial knowledge of a curve's parametric structure, for example, that mixture representation is appropriate, or that some basic model g and some data transformation v is appropriate, can be used to tease apart the curves that represent aspects of a distribution. These individualized curves can in turn be used to make decisions concerning one aspect of distributional structure that is unconfounded with other aspects of that structure, for example, with the exact value of C.

Section 6.2 will consider the similarity between individualized curves like g and v, and the individual orthogonal functions that comprise the Fourier series representation of a pdf or cdf. However, there is a major point of departure between the orthogonal and the elementary function model individualization processes. The orthogonal components that are combined to form a generalized curve representation are solely means to an end, the estimated curve. On the other hand, the individualized curves discussed in this and the next three chapters are in themselves of value and are not merely parts of a whole. This idea is a central point of distinction between the parametric and nonparametric approach. Like the Eskimo word for snow, which has many subtle variants, terms such as curve parameters, coefficients, and properties, must be carefully defined and distinguished before the central point of distinction between parametric and nonparametric approaches can be discussed meaningfully. Hence, this chapter will be concerned primarily with these distinctions.

Many of the distinctions discussed in this chapter involve two concepts, for example, exploratory and confirmatory methods, models and extended models, and parameters and properties. In particular, as discussed in Chapter 2, the constant (1/2) within the normal curve's exponent illustrates that models are sometimes defined in a manner that serves to camouflage distinctions between important concepts, such as that of curve property and curve parameter.

Particular curves will be discussed in Section 6.2 and in many other sections of this book that are closely related to the fractile diagram. The fractile diagram has often been used to modify simple curve estimators, such as the cumulative frequency polygon, and then to display the results of the curve estimation process by means of special graph paper (Hald, 1952, Chapters 5 and 6). When done by hand, this is a tedious process. Today's computers have, however, all but eliminated the need for this form of implementation, together with the lognormal probability paper and other hand calculation aids by means of which fractile diagrams and many closely related curves were constructed. Modern computation also makes it possible to construct and estimate many modifications of these closely related curves, such as the Q-Q plot.

Below, f and F represent the density and cumulative distribution function (cdf), respectively, of some iid sample's elements, g represents some distribution's density model, and G^{-1} represents this same distribution's one-to-one inverse cdf. Also, for the purposes of this book, the Q-Q plot will be treated as an estimator of the curve $G^{-1}F(x)$. Besides $G^{-1}F(x)$ itself, its derivative, the generalized failure rate function, $r(x) = f(x)/g[G^{-1}F(x)]$ (Barlow and Van Zwet 1970; Singpurwalla and Wong 1983, Section 7) and the reciprocal of $r(x)$, $R(x) = g[G^{-1}F(x)]/f(x)$ (Mickey et al. 1963; Tarter and Kowalski 1972; Tarter and Freeman 1987; Tarter 1976; Tarter 1978b; Parzen 1979) can be effectively used to interface generalized and model-based methodology to deal with what R.A. Fisher (1922) called "Problems of Specification."

Specification issues motivate exploratory curve estimation applications which differ from applications described in previous chapters. For example, it will be illustrated in Sections 6.3 and 6.4 that like $G^{-1}F(x)$ the curves $r(x)$ and $R(x)$ do not, in themselves, serve as the end products of statistical analyses. Instead, they are "exploratory tools," whose purpose is quite different from that served by confirmatory and confidence band methods, which are commonly thought of as methods by which evidence is assessed. Exploratory procedures are designed to gather and examine clues, prior to, or as a supplement to, the assessment of evidence.

The split between confirmatory and exploratory statistical approaches is like the division of forensic and legal processes for the gathering of clues and the assessment of evidence. For example, the goal of many trials, to find a defendent innocent or guilty, resembles the goal of accepting or rejecting some null hypothesis in classical frequentist inference. While both the criminal justice system and frequentist statistical inference are designed to provide a yes or a no answer, something quite different is expected of the $G^{-1}F(x)$, $r(x)$ and $R(x)$ estimators. These are displayed graphically, and in their entirety, in order to disclose some important distributional characteristic by the shape they assume.

The use to which $G^{-1}F(x)$, $r(x)$ and $R(x)$ are put is similar to the application of the mixture component λ^2-variance reduced curves, $f_\lambda(x)$, described in the previous chapter. $G^{-1}F(x)$, $r(x)$ and $R(x)$ are usually conceptualized as being composites of pdfs and cdfs, while $f_\lambda(x)$ describes a single pdf. Nevertheless, all four of these curves disclose vital distributional information by their overall shape. While $f_\lambda(x)$ provides a graphical indication of mixture component curve separation, the straightness of curves related to $G^{-1}F(x)$ can be used to determine transformation and model validity. Just as $f_\lambda(x)$ can disclose where curves should be separated, curves related to $G^{-1}F(x)$ can help determine whether assumptions are valid or invalid.

The individualized curves discussed in this and the next three chapters are in themselves of value, and are not merely parts of a whole, because they can help determine both which, and where, assumptions are valid. This is a central distinguishing characteristic between the parametric and model-free approaches. While the kth term of a Fourier curve representation differs conceptually from the $k+1$st term, it does not do so in a manner that is interpreted as having a meaning in and of itself. On the other hand, a location parameter differs in a far more profound sense, from a scale or skewness parameter, than the kth and $k+1$st generalized representational terms or coefficients differ. This is because through many years of experience, the statistical community has slowly evolved a system by which models can be broken down in a manner that has meaningful consequences to practical problems which concern the goodness-of-model-fit.

The need to consider carefully the goodness-of-fit issue was recognized in the formative years of statistical theory. For example, in a 1903 review of a book by the biologist Johannsen in the journal *Nature* the review's anonymous author protested: ". . . against any such crude process of determining goodness of fit as that of placing a normal curve down on seven or eight blocks forming a 'histogram,' and judging the look of the fit. No such test is valid, and further, he [Johannsen] has not yet shown that the normal curve of errors itself is suited to describe the phenomena referred to. . . ." Rather than being primarily antithetical, certain refinements of the "crude process" used by Johannsen can shed new light on the meaning and applicability of parameters that, according to Parzen (1979), are the central conceptualization of statistical inference.

The advantage of associating individual parameters with individual curves is that just where fit is good, and where it is not good, can be assessed with respect to these curves. This localization process is closely related to other forms of curve estimator localization. While the weighted MISE metric introduced in Chapter 4 assesses inaccuracy as a real valued nonnegative scalar, this scalar does not necessarily deal uniformly with all curve arguments. In this respect, the metric serves a purpose that is substantially different from the purpose served by such conventional approaches as simple linear regression confidence bands. Rather than assessing inaccuracy as a real-valued scalar, these bands are conceptualized as either enclosing or not enclosing the population regression curve. That fit may be tight somewhere and loose somewhere else makes no sense when a curve either is, or is not, within some band. This distinction motivates the composite approaches taken in this and the next chapter, where locally optimized estimators of basic curves such as f and F defined above are used to estimate the more complex $G^{-1}F(x)$, $r(x)$ and $R(x)$ curves.

There is a second distinction between parametric confidence band estimators and estimators of curves such as f, F, $G^{-1}F(x)$, $r(x)$ and $R(x)$. When the evaluation of a pdf curve estimator at a point is surrounded by an interval, this interval is used to assess an estimate of probability density. While probability density at a point x_0, $f(x_0)$, can be interpreted as being a property of a pdf, it requires a reversal of usual statistical thinking to interpret $f(x_0)$ as a continuous, as opposed to a discrete, variate's density's parameter. Hence, as will be elaborated upon in Chapter 8, a curve estimator is not a "point estimator" in the sense that this term is used to refer to the estimated parameter value about which a confidence interval or a confidence band is constructed.

The distinction between model parameters and curve properties is of particular importance for the many applications that involve more than one parameter. Any density model is, of course, functionally related to all its parameter values. But functions of this density model can often be constructed that are only related to a proper subset of these values and, in many cases, are related to only one parameter. The shape of these functions can then often be used individually to obtain information that is specific to that part of the model's goodness of fit which is related to the individual parameter.

Like the question of series stopping rules, the number of parameters within a closed- form expression is an important issue. For this reason, models, such as the two parameter normal, will be distinguished from more complex forms of representation, such as lognormal or Weibull densities, which below will be treated as composites composed of two functions as well as three parameters (Heyde, 1963, Crow and Shimizo, 1988). This will be emphasized by calling elementary functions that contain three or more parameters "extended models." (For example, when G is the extreme value model discussed in the next chapter, the

Weibull extended model cdf $G[\{\log(x-\tau)-\mu\}/\sigma]$ involves the parameters τ, μ and σ, while the Weibull model $G[\{\log(x)-\mu\}/\sigma]$ only involves the parameters μ and σ.)

6.1 Parameters, properties, and multipliers

Fourier series curve multipliers differ from model and extended model parameters. For example, while the multipliers used to define a pdf differ from those of a cdf, as it is commonly defined the pdf and the cdf of the same distribution have the same parameters. Because questions of uniqueness are of central importance to many Fourier series applications, it is useful to apply the term "coefficient" solely as as the designation of a mathematical entity that is either a parameter or a property, in the sense that the numerical value assumed by the parameter can be uniquely defined as being the value of the property. "Fourier coefficients" typically satisfy this definition. As used previously, the term multipliers designate the Fourier coefficients of a kernel that has a convergent Fourier series expansion.

The terms statistic, estimate and estimator designate numerical valued entities that are associated in some way with components of a sample. When applied in the context of statistical problems, this allows the word property to be reserved for characteristics of a population and its associated curves. We will use the words operations and symbols as defined by Dewey, (1938, pages 15 and 51). Because attempts to define parameter without distinguishing a *set* from a *class* tend to be "ambiguous" (Tarter, 1975 page 373), for the purposes of this book the word parameter will merely represent an unspecified numerical value within a closed-form mathematical expression. For example, the representation $y=\theta x$, where $\theta=3$, involves the parameter θ which does not appear in the simple representation $y=3x$. Curves are not numerical values and, to avoid confusion, whenever the symbols f, F, g, G, h, H, or v are used, these individual symbols represent specific curves and not parameters. In particular, the symbols g and G designate what will be called below, standard form curves.

For a cdf F, where $F(x)$ is identically equal to $G[(x-\mu)/\sigma]$, or for a density $f(x)$ that is identically equal to some $g[(x-\mu)/\sigma]\sigma$; μ and σ will be defined to be location and scale parameters, respectively, of $G[(x-\mu)/\sigma]/\sigma$ and $g[(x-\mu)/\sigma]/\sigma$. Consequently, in all instances where a given symbol is said to represent either a location, a scale, a threshold, or any other of the specific parameter discussed below, the use of the terms, location, scale, threshold, etc., will imply that some identity like $F(x) \equiv G[(x-\mu)/\sigma]$ has been previously specified between the curve that, like F or f, describes a variate's distribution and a standard form elementary function G or g. Consequently, the term standard form curve indicates below that a curve like G or g is not specified in terms of any parameter. Note, for example, this implies that the Cauchy density, which is a Student-t density associated with a *specific* number, one, degree of freedom can be a standard form, but that it would be improper to specify that the Student-t density with an unspecified number of degrees of freedom as a standard form. Hence, in an important sense, the standard form g is to the parametric approach what any particular orthogonal function is to series-form generalized representation. Both are completely specified building blocks from which more complex curves are constructed.

Unfortunately, the word "parameter" is phonetically so similar to the word "perimeter" that such phrases as, parameters of the problem at hand, and parameters of the agreement, sometimes are applied to designate the scope, boundary or frame of a problem or agreement. However, parameters encountered in data analysis applications perform a more specific function than do vernacular parameters. In particular, it is useful to clearly distinguish the term *parameter* from the term *property*. This distinction is important because it underscores the point that although a property, like the expectation of an X variate, $E(X)$, can often be designated numerically equal to a parameter like the location parameter μ, there may be many other alternative properties that share $E(X)$'s connection with the symbol μ within some expression for a curve.

That model parameters do not have the same unambiguous connection with curve properties that Fourier coefficients have with the properties $E[\exp(-2\pi ikX)]$, $k=0,\pm1,\pm2,\ldots$, can be both a disadvantage and an advantage. For example, we will show below that a major advantage of this ambiguity is that a single parameter may be associated with all the evaluations of a specific curve. The example of such a curve which is of most relevance here, is the curve $R(x)$, which, for the scale parameter σ, yields alternatives to the curve property, the population standard deviation $\sqrt{\text{Var}(X)} = \sqrt{E\{X-E(X)\}^2}$, with which σ is usually identified. For any one-to-one cdf G there are an infinite number of distinct curve properties (specifically, evaluations of $R(x)$) which are numerically equal to any scale parameter of a cdf F whose standard form is G. Thus, in a general sense, evaluations of $R(x)$ are curve properties, which, when the appropriate standard model g is selected, are identically equal to the scale parameter of the density f's representation in terms of g.

The advantage served by identifying σ with the entire curve $R(x)$, rather than the single property $\sqrt{\mathrm{Var}(X)}$, stems from the role that the standard form G plays as part of the definition of R. A second function, the transformation v, can be inserted as part of the representation of F so that choices can be made about the representation of F in terms of the standard model G or in terms of an extended model based on both G and v. Note, this process is very much like the use of a stopping rule procedure for choosing between a curve estimator that includes, or one that does not include, some $k+1$st term. Here, within the generalized representation, the $k+1$st term plays the same role that the function v plays for the extended model.

We now come to the basic approach that is central to this and the next chapter and, in addition, motivates much current and past curve estimation research. Questions such as the inclusion of some $k+1$st term are typically much easier to deal with within the framework of generalized than in the framework of elementary-function based expression. However, for historical and many other reasons, model-based methods form the central core of statistical thinking. Consequently, it makes sense to use the advantages of approaches based on generalized representation to first estimate f and F and, only then, form these estimators into composites like $G^{-1}F(x)$, $r(x)$ and $R(x)$ which can handle the specification issues that greatly limit the effectiveness of model-based methodology.

Applications based on $G^{-1}F(x)$, $r(x)$ and $R(x)$ utilize the fact that density and cumulative models as well as extended models parameters are symbolic representations that serve to express some curve or function, like f, in terms of some standard form curve or function, like g. Thus, for example, when it does not connect some f with some g, a symbol that represents skewness is not a parameter. For both models and extended models, parameters can be distinguished from properties in terms of the roles they play in connecting f with g. These roles can be studied in a manner similar to the determination of whether the nouns that compose a sentence serve as subjects, objects or other parts of some sentence.

A scale parameter σ could not suggest *how* a density is spread along the x-axis without the division slash with which it is prefixed. Since like the object of an English sentence, the location-ness or scale-ness of a location or scale parameter arises from its position and adjoining symbols in the extended model, that the internal structure of the extended model contains certain parameters makes it possible to turn the extended model inside out; a process by which individual parameters can be identified with curves that, when a representational choice is valid, will take on easily recognized shape.

In contrast to parameters, curve *properties* are defined extrinsically. For example, the mean E(x) is a not a specific curve property of some density or cdf because of the arrangement of symbols within a form of mathematical shorthand like $G[(x-\mu)/\sigma]$. Instead E(x) abstracts the information conveyed by a certain curve or function. As is well known, the mean and the standard deviation are usually associated with the location and scale *parameters* μ and σ, but can (for some models) either be ambiguously defined or (in the case of the standard deviation of the Cauchy with finite scale parameter σ) be infinitely valued.

An individual *property* plays the role of a one-word abstract of a sentence. For this reason we will use the word property to designate the result obtained when a series of well-defined numerical operations, such as integration procedures used to compute E(X) or numerical procedures used to evaluate $R(x)$, are applied to some curve representation whose parameters are specified. Two properties will be said to be distinct when the specific operations used to determine the value assumed by the properties can produce different numerical values from the same curve. It follows that the term "nonparametric" should be reserved as a designation for estimates based or targeted exclusively on properties of a curve, such as the mode, Parzen (1962).

Properties will be said to "not exist" when either they are ambiguously defined by some specified operations or when the application of these operations will produce a value that equals or approaches infinity. Property existence is, of course, particularly a problem associated with the derivatives of a function evaluated at a point or definite integrals whose limits of integration approach infinity in an incompletely defined fashion, or, even when defined more completely, when the function has fat tails that extend to infinity in either or both directions.

Generally speaking, the properties of a curve are highly curve-specific. On the other hand, if scale and location *parameters* μ and σ of $F(x) = G[(x-\mu)/\sigma]$ are known, this knowledge discloses something about G's relationship to F, *no matter what cdf F actually is*. This knowledge, which does not correspond to any application of any properties of a function, can be utilized to guide and then perform the computations that underlie certain basic statistical processes. For example, the values assumed by the cdf $F(x)$ along the x axis are by the above definitions properties, and not parameters, of this curve. Hence, the use of these properties to study the structure of F would be defined as being a nonparametric procedure.

Suppose, that the pdf f, and some underlying model g, have one-to-one cdfs F and G, where G^{-1} represents the inverse function of G. Suppose further that $\mu=0$ while scale parameter σ is unspecified. But, in reality, is σ truly unspecified? Consider that by the definition of the term *scale* parameter, the identity $\sigma \equiv x/G^{-1}F(x)$ must apply for *any value* of the running variable x. Thus, defining σ to be a scale parameter is equivalent to the assertion that the identity $\sigma \equiv x/G^{-1}F(x)$ is valid. This, in turn, implies that a particular composite function, $x/G^{-1}F(x)$ is a horizontal line. Another way of phrasing this assumption is that the horizontal linearity of the curve $x/G^{-1}F(x)$ is a necessary condition for σ to be the scale parameter of F with respect to underlying G. Like the shape assumed by the $x/G^{-1}F(x)$ function, the shape of the $R(x)$ function depends on an assumed underlying model G. An estimator of $R(x)$ combines this choice with estimators of f and F to both: (1) detect the degree and the nature of model goodness-of-fit; and, (2) when the fit of a model is found to be poor, help determine the most appropriate choice of model or extended model alternative.

The curve $R(x)$ differs from the curve $x/G^{-1}F(x)$ because it can be used to answer questions pertaining to two functions G and v rather than to merely the single function G. Estimates of $R(x)$ can in turn be based on two distinct sets of curve properties, the evaluations of cdf F, and the evaluations of pdf f. Correspondingly, two sets of multipliers are used to estimate a single $R(x)$ curve by the following process: first, members of the sequence $\{\hat{B}_k\}$ are calculated; then $\{\hat{B}_k b_k^{(1)}\}$ and $\{\hat{B}_k b_k^{(2)}\}$ are determined, f and F are estimated and, finally, $R(x)$ is constructed.

The advantage of the above complex approach stems from the dual nature of the specification process. An investigator is often as unlikely to be sure of G as he or she is sure of v. For example, a commonly used extended model is the special case of $G\{(\mathrm{v}[x-\tau]-\mu)/\sigma\}$, where for the constant $K \neq -1$, $\mathrm{v}(z)$ is chosen to equal $z^{[1/(1+K)]}$, which below will be called a fractional power extended model. Below we will use the term logmodel to designate the cdf $G\{(\log[x-\tau]-\mu)/\sigma\}$, and since we will no longer explicitly discuss kernels, we will use the symbol K to represent part of the exponent of $z^{[1/(1+K)]}$. Logmodels are the often applied limiting cases of fractional power extended models. At any x, for any finite value of $K \neq -1$ the extended model scale parameter is numerically equal to evaluations of the curve $[R(x)]^{[(1+K)/K]}$ (which for the square root transformation special case is $[R(x)]^2$ and for the reciprocal transformation case is $[R(x)]^{1/2}$).

$[R(x)]^{[(1+K)/K]}$ curves, where $K \neq -1$, are to the extended model $G\{(\mathrm{v}[x-\tau]-\mu)/\sigma\}$, where $\mathrm{v}(z)=z^{[1/(1+K)]}$, what $x/G^{-1}F(x)$ is to the model $G[(x-\mu)/\sigma]$. The $R(x)$ curve is also to $G[\{\log(x-\tau)-\mu\}/\sigma]$ what $x/G^{-1}F(x)$ is to $G[(x-\mu)/\sigma]$. Specifically the estimated straightness and slope of $R(x)$ or $[R(x)]^{[(1+K)/K]}$ can be used to determine: (1) whether certain basic statistical assumptions implicit in the choice of the functions G and v are satisfied and (2) the steps that should be taken if these assumptions are not satisfied.

Since it disentangles extended model components without prior knowledge or estimates of extended model parameters, the application of parsing processes based on $R(x)$ can bypass what Fisher called the estimation phase of the statistical process. Hence, it helps avoid the circular reasoning where verification of the specification phase of statistical analyses requires estimators that are appropriate only if valid decisions are made in the specification phase.

6.2 Coefficients as properties and parameters

Charlier and Wicksell (1924) characterized Pearson's method of moments approach to the decomposition of mixtures as a "heroic effort." Charlier and Wicksell's characterization illustrates the difficulty encountered when forms of parametric representation are used that contain multiple parameters. Difficulties with the complex relationships between multiple parameters and multiple properties is much more pronounced for mixtures, extended models and mixtures of extended models than it is for models. For example, the properties mean $\mathrm{E}(X)$, and standard deviation $\sqrt{\mathrm{Var}(X)}$ do not have a simple relationship with the parameters specified within $G\{(\mathrm{v}[x-\tau]-\mu)/\sigma\}$ for any non-identity function v. However, unlike forms of curve representation such as $G\{(\mathrm{v}[x-\tau]-\mu)/\sigma\}$, the forms of representation emphasized in this book, Fourier expansions, have coefficients that can, in all but a few circumstances, be unambiguously interpreted from both the point of view of parameters and of properties. The first tool that was devised to make this correspondence between density parameters and Fourier expansion properties perfect was the Fejér multiplier sequence. Hence, even in the rare circumstances where a Fourier expansion of a density does not correspond without emendation to that density, the Fejér approach, which is strikingly similar to what was previously introduced as the lambda method, can be used to remove all practical difficulties.

A function or curve H represented as a Fourier series contains coefficients that play precisely the same role for this curve that the parameters τ, μ and σ play for $G\{(\mathrm{v}[x-\tau]-\mu)/\sigma\}$. On the other hand, each of the Fourier coefficients of H, B_k, where $-\infty<k<\infty$, is represented in a form similar to the expression for the kth moment of a density about the origin, $E(X^k)$. Unlike moments however, for almost all practical purposes the existence and applicability of sequences composed of individual B_k terms is not a critical issue.

The near identity between Fourier coefficients as parameters and these same entities as properties greatly simplifies many statistical applications. Fourier representation also makes it possible to utilize multiplier sequences that can determine: (1) which specific form of curve, for example a pdf or a cdf, is selected or (2) the manner by which a curve estimator is optimized with respect to the curve it estimates.

Because it is so central to the methodology described in this book, it should be mentioned that in many circumstances separate multiplier sequences that determine (1) and (2) can be combined to form a single multipurpose sequence. Functions that can be represented as composites or compounds of other simpler curves can take both computational and statistical advantage of particular choices of multiplier sequences. Here the term statistical advantage refers to the efficient use of available data for the accomplishment of a specific purpose. For example, if it is important to tease apart a mixture density into its component parts, this process can be performed sequentially. First, a multiplier sequence that uses data efficiently for the purpose of estimating the right side of a density can be applied to the overall estimator. Then the right side of the density can be decomposed, isolated, and separated. Finally, the same process can be repeated for the left density side, using an implementation strategy that enhances data efficiency for the purpose of estimating the left side.

6.3 Curves and multiple parameters

The derivation of the identity $\sigma \equiv x/G^{-1}F(x)$ required the assumption that the parameter μ equals 0. But, for the two parameter model

$F(x) \equiv G[(x-\mu)/\sigma]$, $G^{-1}F(x)$ is identically equal to $(x-\mu)/\sigma$, a two—not a one—parameter function. However, for differentiable G^{-1} and F, the derivative of $G^{-1}F(x)$ is identically equal to $1/\sigma$. As will be illustrated below in this, and in Chapters 7 and 9, differentiation can be used to individualize parameters other than σ.

It is informative to note that the derivative of $G^{-1}F(x)$ is a curve which shares a crucial feature possessed by the isolated mixture component curves discussed in the previous chapter. Isolated mixture component curves are only related to two of the five parameters of the two-component mixture model. Similarly, the derivative of $G^{-1}F(x)$ is only related to the single parameter σ. In the following sense, when the reciprocal of the derivative of $G^{-1}F(x)$ is identically equal to the single parameter σ, this identity connects certain properties of F with a single parameter of the model-based representation of F: If the model cdf G is a valid choice, then when evaluated at any point x_0, the property, reciprocal of the derivative of $G^{-1}F(x)$, equals the parameter σ no matter what value is assumed by the parameter μ. In a similar fashion, the kth Fourier coefficient of a density f will equal $E[\exp(-2\pi ikX)]$ no matter what values are assumed by the other Fourier coefficients of f's expansion.

With the notable exception of Fourier coefficients, multiple distinct properties can conveniently be associated with a specific parameter. We will see in the next chapters that this fact can be turned to advantage and curves like the reciprocal of the derivative of $G^{-1}F(x)$ can be constructed that, in certain circumstances, are identically equal to parameter values. In other circumstances these curves can provide graphical clues, not only that an incorrect model has been assumed, but, when assumptions are incorrect, these curves can provide useful clues concerning remedial procedures. In both the univariate and bivariate cases these curves exist for not only the scale parameter σ, but for what is commonly called the threshold parameter τ, as well as the parameter identically equal to the K of the fractional power extended model $G\{(v[x-\tau]-\mu)/\sigma\}$, where for $K \neq -1$, $v(z)=z^{[1/(1+K)]}$. That the parameters of a curve take on a specific value implies that this specific value is unrelated to other curve parameters as well as to the argument at which the curve is evaluated. Below, we will make considerable use of this characteristic of model and extended model parameters.

One reason why this characteristic is of value is illustrated by an observation about the lognormal extended model made by Cohen (1951) as follows:

> The logarithmic normal distribution provides a useful theoretical model for studying a number of biological populations, certain economic populations involving income distributions, and others in which the standard deviation of individual observations is approximately proportional to the magnitude of the observations.

For any one-to-one model G, what Cohen refers to as the property, standard deviation, can be interpreted in terms of the parameter σ of any logmodel $G\{(\mathrm{v}[x-\tau]-\mu)/\sigma\}$, where $\mathrm{v}(z)=\log(z)$. Additionally when $K \neq -1$, and $\mathrm{v}(z)=z^{[1/(1+K)]}$, the parameter σ of the extended model fractional power $G\{(\mathrm{v}[x-\tau]-\mu)/\sigma\}$ can be associated with the linearity of the curve $[R(x)]^{[(1+K)/K]}$, yet is functionally unrelated to the parameter μ. This later characteristic is of particular importance in light of Fisher's separation of statistical processes into problems of estimation and problems of specification. Because the curve $[R(x)]^{[(1+K)/K]}$ is functionally unrelated to the parameter μ, it is safe to use this curve to deal with traditional specification issues without worry that the resolution of these issues will be confounded with issues involving the traditional processes of parameter μ's estimation.

Cohen's lognormal characterization concerning the approximate proportionality to the "magnitude of the observations" implies a form of linear relationship or curve. That this curve is linear helps interpret lognormal goodness-of-fit in the same way that the linearity of other curves can help deal with specification issues involving many other models and extended models. Fortunately, there is no need for the word approximation here, as in Cohen's characterization of the lognormal. Quite, to the contrary, there are many useful exact relationships between curves and parameters.

6.4 An example of the simultaneous selection of a model and a transformation

The six curves superimposed on the estimated density shown in Figure 6.1 illustrate how an exploratory process that involves two curves can be extended to include a last step which facilitates the choice between alternative models and transformations. The bell shaped curve estimator of f shown in this figure was obtained by applying the procedures described in Chapter 4 for a choice of MISE weight function $w(x)=1$, to 1000 iid N(3,1) variates.

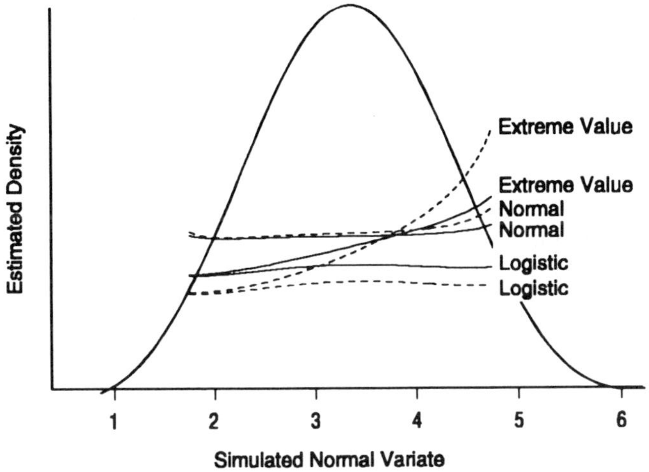

Figure 6.1. *Estimated component of marginal and superimposed $R(x)$ curves. The solid curves are the $R(x)$ functions that correspond to log transformations, while the dashed curves correspond to square root transformations.*

Based on the bell-shaped estimator of f, the six $R(x)$ curves shown in Figure 6.1 correspond to combinations of three models, the normal, extreme value, and logistic; and two transformations, the log and square root. Each curve can be used to assess the fit of one model combined with the choice of one transformation. For example, the straightness of the bottom curve can be used to judge how well the logistic model would fit the data after an initial square root transformation had been applied.

That the extreme value - Weibull $R(x)$ curves differ so substantially from the normal or logistic $R(x)$ curves is highly informative. The extreme value density is not symmetric. However, the three parameter Weibull, which is to the extreme value what the lognormal density is to the normal, can fit most symmetric densities very well. However, in the case shown, that neither of the extreme value Weibull $R(x)$ curves is a straight line indicates that, despite the great flexibility of the three parameter Weibull curve, it would be foolish to assert that the density f of the distribution is adequately represented by a Weibull extended model, or alternatively, assert that f should be represented as an extended model with square root v and extreme value density g components. Instead, the fact that all four curves besides the extreme value curves are fairly straight yet horizontal, indicates that either the two parameter normal or the two parameter logistic would be a better choice than would either the two parameter extreme value model or even the three parameter Weibull extended model.

That the four curves other than the extreme value curves are all fairly horizontal indicates that neither the log nor the square root transformation would improve the fit of either the normal or the logistic model.

We will now proceed to discuss the technical details that underlie the construction of $R(x)$ and related curves.

CHAPTER 7

Threshold parameter and transformation applications

In this chapter the distinctions and definitions discussed in Chapter 6 will be applied to formulate systematic approaches to model and transformation selection based on generalized forms of representation. Because of its direct relevance to both threshold parameter τ determination, as well as transformation selection, emphasis will be placed on the curve $R(x)$ as the vehicle through which individual parameters will be associated with individual curves. We will also describe how the straightness and slope of $R(x)$ can determine:

(1) whether certain basic statistical assumptions implicit in the choice of the model G and transformation v are satisfied; and
(2) if these assumptions are not satisfied, what steps should be taken.

We will also show that the fraction with numerator $R(x)$ and denominator $R'(x)$ can be used to determine the value of the constant K, where $K \neq -1$, of the fractional power transformation $\mathrm{v}(z)=z^{[1/(1+K)]}$ without the need for preliminary estimates of the parameters μ_1, μ_2 and σ of what we will call the v-model $G\{(\mathrm{v}[x-\mu_1]-\mu_2)/\sigma\}$. In many, but as explained below not all applications, μ_1 is identical to what Crow and Shimizu (1988) call the "threshold parameter τ." Curves like $R(x)$ and $R(x)/R'(x)$ differ substantially from curves generated by applying of the lambda mixture decomposition procedures described in Chapter 5. Hence, before describing $R(x)$ curve implementation we will compare this curve's role in the parsing of v-models to the role of lambda contrast enhanced curves in the decomposition of mixtures like $f(x|p,\mu_1,\mu_2,\sigma) = \{pf_1[(x-\mu_1)/\sigma] + (1-p)f_2[(x-\mu_2)/\sigma]\}/\sigma$. It is also useful to compare the $R(x)$ curve to a curve that is to the association between two variables what $R(x)$ is to the parameter σ of a univariate distribution.

Rietz and Crathorne (1924, pages 129–132) describe a useful statistical tool called the "correlation ratio" represented by $r^2 = 1 - s_y^{\,2}/\sigma_y^{\,2}$. Tarter (1978a); Doksum, Blyth, Bradlow, Meng (1991); Doksum, Blyth, Bradlow, Meng and Zhao (1992); Doksum and Samarov (1991), and Bjerve and Doksum (1992) have discussed the estimation and applications of the correlation ratio treated as a curve. It is not obvious from Rietz and Crathorne's notation how the correlation ratio can be viewed as a curve. Using updated statistical notation the correlation ratio can be conceptualized as the function $\rho(x) = \sqrt{1 - \sigma_{y|x}^2/\sigma_y^2}$, where $\rho(x)$ is the square root of a quantity equal to one minus a fraction whose numerator is the variance of the conditional density $f(y|x)$ of a Y or response variate given the value x of an X or key variate. The denominator of this fraction is the variance of the marginal density of the response variate. If the variance of $f(y|x)$ is constant for all values of x, then $\rho(x) \equiv |\rho_0|$, the absolute value of the usual Pearson correlation coefficient ρ. That $\rho(x) \equiv |\rho_0|$, symbolically designates that a correlation ratio curve is a horizontal line, specifically, $y = |\rho_0|$.

Suppose that the property of the bivariate density $f(x,y)$ with x and y means μ_x, and μ_y respectively, the population correlation coefficient

$$\rho = \frac{E_f\left\{(X-\mu_x)(Y-\mu_y)\right\}}{\sqrt{E_f\left\{(X-\mu_x)^2\right\}E_f\left\{(Y-\mu_y)^2\right\}}},$$

can be unambiguously identified with a single parameter within some elementary function-based expression for $f(x,y)$. Below, this parameter will be called an association parameter. Based on the assumption that there is a single association parameter, which we will call the null case, the function $\rho(x)$ is the curve identically equal to this association parameter. On the other hand, in the nonnull case the nonhorizontal or nonlinear shape of $\rho(x)$ will disclose that it is simplistic to subsume the dependence of the response variate Y on the key variate X by a single number. In this way $\rho(x)$ can help shed light on the complex structure of the dependence relationship. For example, suppose $\rho(x)$ is horizontal for small and moderate values of $x < x_0$ and then, at and beyond x_0, it changes in form to a line of positive slope. In this situation the shape assumed by $\rho(x)$ indicates that the degree of relationship between two variates is constant for x less than x_0 but that it changes beyond x_0; beyond x_0, x it determines or is associated with a steadily increasing portion of the response variate y.

The above application of $\rho(x)$ illustrates the following two important points:

(1) the uniqueness of a particular property of a distribution, like the unique value assumed by a particular parameter of a model or extended model, can be interpreted as a particular curve being a horizontal line.

(2) departures from an assumption can be gauged by the shape of a curve that in the null case would be a horizontal line, but in the nonnull case cannot be a horizontal line.

It is informative to compare the $\rho(x)$ and $R(x)$ curves to the lambda contrast enhanced curve $f_\lambda(x)$ discussed in Chapter 5. All three curves share the following characteristic. Their geometric curve features provide the information needed to discern some important distribution attribute. In the case of the $\rho(x)$ curve, the attribute in question is whether or not a single property, the correlation coefficient ρ, encapsulates information concerning the interdependence of two variates. When $\rho(x)$ is not a horizontal line, no single number completely determines exactly how two variates are related to each other.

In keeping with the perspective introduced in the previous chapter, there exists a single association parameter that is unambiguously connected to a single curve property ρ_0 only when the curve $\rho(x)$ is identically equal to ρ_0. Also, just as in the case of a model's scale parameter σ, there is no necessary reason why some commonly used curve property, for example, the correlation coefficient, need be the only curve property whose value equals the value of a parameter like ρ within a bivariate model's representation. For instance, the same simulation methods used to generate correlated normal variates can be easily adapted to simulate dependent Cauchy variates. Specifically, Scheuer and Stoller (1962) give an example of a set of three linear equations by which three independent N(0,1) variates can be used to generate a trivariate vector of normal variates with a specified variance covariance matrix. For the bivariate case, within the Scheuer Stoller algorithm used to generate correlated normal variates is a user setting ρ which inputs the desired correlation between normal variates. Yet this same setting cannot be unambiguously associated with the *property* correlation for dependent Cauchy variates because

$$\mathrm{E}_f\{(X-\mu_x)^2\}=\mathrm{E}_f\{(Y-\mu_y)^2\}=\infty$$

for the univariate Cauchy variates X and Y used to generate dependent Cauchy variates.

Note that the Scheuer Stoller linear equations can be easily adapted by substituting standard Cauchy variates, where the standard Cauchy density g was defined in Chapter 5, for N(0,1) variates. In this case, the symbols that represented covariances and correlations in the Scheuer Stoller equations cannot represent these properties, since these properties do not exist for the Cauchy. Clearly then, when a property of dependent Cauchy variates' bivariate distribution equals the finite ρ setting within the Scheuer Stoller equations this property cannot be

$$\mathrm{E}_f\{(X-\mu_x)(Y-\mu_y)\}\Big/\sqrt{\mathrm{E}_f\{(X-\mu_x)^2\}\mathrm{E}_f\{(Y-\mu_y)^2\}}.$$

The Scheuer Stoller bivariate Cauchy illustrates that a given parameter of a model or extended model's closed form representation can be numerically equal to a property other than that property which, like the correlation coefficient, is usually associated with the parameter. Until the last section of this chapter we will take advantage of this latitude of choices in the univariate case. Nevertheless, in this and the next two chapters we will deal with a question of relevance to applied statistical problems irrespective of curve dimensionality: From among the many curve properties that can be made to correspond to components within a model's, or extended model's, parametric representation, how can certain specific properties be used to construct informative graphical tools?

The curve $R(x)$ described below is an example of one answer to this question because, among other applications, it can help choose between alternative transformations v and alternative univariate models G within the expression for a cdf, $G\{(\mathrm{v}[x-\mu_1]-\mu_2)/\sigma\}$. From a general curve estimation perspective, the single curve $R(x)$, constructed as a composite of the two functions f and F, can be used to make decisions about the two curves G and v. One particularly important candidate choice of v is the identity function $\mathrm{v}(x)\equiv x$. Thus, in the same way that the $f_\lambda(x)$ curve can help make decisions about the appropriateness of the mixture form of the extended model, the $R(x)$ curve can serve as a graphical tool for the analysis of the extended model $G\{(\mathrm{v}[x-\mu_1]-\mu_2)/\sigma\}$ and its $\mathrm{v}(x)\equiv x$ special case, the model $G[(x-\mu)/\sigma]$.

When the mixing parameter p satisfies the inequalities $0<p<1$, the $f_\lambda(x)$ curve can help discern the structure and estimate the parameters of representational components. In certain instances where $f_\lambda(x)$ can be based on a large sample of data from a mixture of v-models whose components overlap only moderately, $f_\lambda(x)$ can be used to separate individual components, that can then be studied by means of $R(x)$-based graphical analysis. Thus, in one sense, $R(x)$ and $f_\lambda(x)$ are curve estimation alternatives and, in a second sense, they can be viewed as complementary approaches. A data analyst commonly must decide whether or not a skewed curve results from some mixture of highly overlapped components or, alternatively, if a single model will represent the density in question once an appropriate transformation is determined. He or she may even find that some mixture decomposition process separates components that individually can only be modeled by the lognormal or some other extended model, rather than some model proper.

Application of the $R(x)$ curve, like use of the curve $f_\lambda(x)$, requires some user control. Specifically, $R(x)$ is based on candidate choices of G and v. While $f_\lambda(x)$'s informativeness depends on the assumptions: (1) that the property, second cumulant, is associated with the parameters $(\sigma_s^2-\lambda^2)$, $s=1,\ldots,c$; and (2) that changes in the value selected for λ of $(\sigma_s^2-\lambda^2)$ do not alter the mixing and location parameters of the density's representation, $R(x)$'s effectiveness is tied to the validity of the extended model $G\{(\mathrm{v}[x-\mu_1]-\mu_2)/\sigma\}$. However, $R(x)$ is functionally independent of one v-model parameter, μ_2, while $R(x)$'s slope and y-intercept are related to the other two parameters, μ_1 and σ. We will also show in the next section that $R(x)$'s slope can be used to formulate a parametric approach to the property skewness.

Sections 7.2 through 7.5 will deal with the specifics that underlie $R(x)$-based methods. For example, useful relationships will be derived for the special cases of the v-model, where v within $G\{(\mathrm{v}[x-\tau]-\mu_2)/\sigma^*\}$ is the log function, which we will refer to as logmodels. It will be shown that as $\tau\to-\infty$, when $\sigma^*=|\sigma/\tau|$ and $\mu_2=[\log(|\tau|)+\mu/\sigma]$, any logmodel approaches the model $G[(x-\mu)/\sigma]$. This limit relationship corresponds to the fact that a curve $R(x)$ which is a horizontal line implies that some model $G[(x-\mu)/\sigma]$ provides a valid representation of cdf F, while an $R(x)$ which is a diagonal line with slope ϵ implies that logmodel representation is appropriate. That as $\epsilon\to0$ the diagonal $R(x)$ line approaches a horizontal line, corresponds to model $G[(x-\mu)/\sigma]$ being the $\tau\to-\infty$ limiting case of a logmodel $G\{(\log[x-\tau]-\mu_2)/\sigma\}$.

Treating models as limiting cases of their associated logmodels helps determine whether or not it is worthwhile to assume a three-parameter logmodel, like certain Weibull or lognormal extended models, in place of some corresponding two-parameter model, like the extreme-value or normal model. Once the choice of model or extended model is determined by judgments concerning the straightness and slant of $R(x)$, the functional connection between this curve and v-model parameters can be used for many purposes. For logmodels and fractional power choices of v, the slope and y-intercept of a line fitted to R can be applied to estimate the two parameters, μ_1 and σ. Since $R(x)$ is functionally independent of μ_2's value, this parameter cannot be estimated directly using $R(x)$. However, once two of the three constituent parameters of the v-model are estimated, it is almost always the case that the remaining parameter can be effectively estimated using conventional methods. For example, consider the logmodel special case of the v-model. Once the value of the parameter μ_1 within $\log(Y-\mu_1)$ is determined, $\log(Y-\mu_1)$ can be applied as a data transformation. Assuming that G has been determined by graphing $R(x)$, or in some other way, conventional location parameter estimates of μ_2 based on this choice of G can be obtained from this sample of post-transformation data.

The methods described both in this chapter and in Chapter 5 can be applied to v, G, f_1 and f_2, which are not elementary functions. However, examples of the extended model $G\{(\mathrm{v}[x-\mu_1]-\mu_2)/\sigma\}$ and the extended model $f(x|p,\mu_1,\mu_2,\sigma)$ usually encountered in the statistical literature are comprised of two distinct elementary functions and three parameters. Despite this similarity, the $R(x)$ curve used to parse or decompose $G\{(\mathrm{v}[x-\mu_1]-\mu_2)/\sigma\}$ and the $f_\lambda(x)$ curve used to decompose $f(x|p,\mu_1,\mu_2,\sigma)$ are defined and applied in very different ways. Consequently, before turning to the details of $G\{(\mathrm{v}[x-\mu_1]-\mu_2)/\sigma\}$ decomposition, it seems worthwhile to discuss why it is that parsing the v-model class of extended models differs substantially from the decomposition of finite mixtures.

The key structural feature that differentiates $G\{(\mathrm{v}[x-\mu_1]-\mu_2)/\sigma\}$ from $f(x|p,\mu_1,\mu_2,\sigma)$ is the placement of the function v as well as the parameters μ_1, μ_2 and σ within the argument of the function G. The advantage of this arrangement can be seen by considering the identity $F(x)\equiv G\{(\mathrm{v}[x-\mu_1]-\mu_2)/\sigma\}$. When G is a one-to-one function, this implies the existence of an inverse function of G, G^{-1}, and, as a consequence, the validity of the identities

$$G^{-1}F(x) \equiv G^{-1}G\{(\text{v}[x-\mu_1]-\mu_2)/\sigma\}$$
$$\equiv (\text{v}[x-\mu_1]-\mu_2)/\sigma.$$

When G takes on some assumed form, the left side of these identities, $G^{-1}F(x)$, is a curve that can be estimated from sampled data using procedures discussed in Chapter 3 to estimate F. Consequently, when v can be assumed to be either the log function or a fractional power function, the right side of the identity contains a function v that has a recognizable derivative. Specifically, (1) for the log special case of v, the curve $R(x)$, and (2) for fractional power v, the curve which is some known power of $R(x)$, will be a straight lines.

In addition, when the derivatives of F and G, f and g, as well as F and G's second derivatives, f' and g', are well defined, and when v is a fractional power function then

$$R(x)/R'(x) \equiv \frac{f(x)}{g'G^{-1}F(x)/R^2(x) - f'(x)}$$

will be a straight line. Unlike the argument G of $G\{(\text{v}[x-\mu_1]-\mu_2)/\sigma\}$, neither f_1 nor f_2 within the representation of $f(x|p\mu_1,\mu_2,\sigma)$, has the other function within its argument. It is additivity with respect to the two components, $pf_1[(x-\mu_1]/\sigma)/\sigma$ and $(1-p)f_2([x-\mu_2]/\sigma)/\sigma$, that allows the curve $f_\lambda(x)$ to expedite the mixture decomposition process. This additivity permits the lambda multiplier sequence $\{b_k^{(\lambda)} = \exp[2\pi^2k^2\lambda^2]\}$ to distribute among the individual components of $f(x|p\mu_1,\mu_2,\sigma)$. Because of the association between choices of λ and the second cumulants of all components of the estimated mixture, when both f_1 and f_2 are symmetric about zero, and the location parameters μ_1 and μ_2 are sufficiently separated, lambda contrast enhancement clarifies the structure of the mixture. The lambda process generates one curve, $f_\lambda(x)$, for each multiplier sequence $\{b_k^{(\lambda)} = \exp[2\pi^2k^2\lambda^2]\}$. $R(x)$ is the reciprocal of the *derivative* of $G^{-1}F(x)$ and can be estimated through the application of two multiplier sequences.

Consider the derivative of function $G^{-1}F(x)$. This derivative involves both the function F itself and the derivative of the function $F' \equiv f$. It is for this reason that two distinct multiplier sequences are involved in the determination of $R(x)$. As described in Chapter 3, a cumulative, here the function F of $G^{-1}F(x)$, can be estimated by substituting a particular multiplier sequence $\{b^{(2)}\}$ for the multiplier sequence $\{b^{(1)}\}$ associated with the density f. Thus, unlike the construction of a single curve $f_\lambda(x)$, two multiplier sequences are used to construct the single curve $R(x)$, specifically $\{b^{(2)}\}$ and $\{b^{(1)}\}$. We also will show in the next section that there are useful applications of the derivative of $R(x)$, $R'(x)$, and that the estimation of $R'(x)$ typically involves three multiplier sequences.

7.1 The $R(x)$ function and skewness

Besides location, scale and association, there are other classes of parameters that characterize either a distribution or some curve associated with a distribution. One of these parameters has a close connection with the function v of the v-model $G\{(\mathrm{v}[x-\mu_1]-\mu_2)/\sigma\}$. Skewness of an estimated density often indicates the need for some data transformtion v. However, before discussing the connection between skewness and the $R(x)$ curve, skewness as a curve property, and skewness as a curve parameter, will be distinguished.

Kendall and Buckland's dictionary (1960) defines skewness as: "An older and less preferable term for asymmetry, in relation to a frequency distribution." Kendall (1952) describes the following three measures of skewness proposed by K. Pearson:

(1) {Mean-Mode}/σ, where for Kendall, σ represents the population standard deviation, which like the mean and mode of the pdf, is a property,

(2) $\Theta = [\sqrt{\beta_1}(\beta_2+3)]/[2(5\beta_2-6\beta_1-9)]$, where $\beta_1 = \mu_3^2/\mu_2^3$, $\beta_2 = \mu_4/\mu_2^2$, and μ_2, μ_3 and μ_4 are respectively the second, third and fourth ordinary population moments about the mean, and

(3) $\gamma_1 \equiv \mu_3/\mu_2^{3/2}$.

It is obvious that even what Kendall refers to as the "more convenient" quantity γ_1 is far more difficult to interpret than are the location and scale parameters μ and σ. Just as the correlation ratio is based on the property *variance* of the conditional density, measures of skewness; {Mean-Mode}/σ, Θ and γ_1, are based on particular *properties* of the specific underlying distribution being considered. They are not defined in a manner similar to the definitions of location and scale. On the other hand, for the $G\{(\log[x-\mu_1]-\mu_2)/\sigma\}$ logmodel special case of the v-model where the density $g=G'$ is a symmetric function, the slope and second derivative of $R(x)$ *are* conceptualizations of skewness and kurtosis which are closely connected with location and scale parameters. For example, suppose $v(z) \equiv \log(z)$. It follows that

$$\begin{aligned} G^{-1}F(x) &\equiv G^{-1}G\{(\log[x-\mu_1]-\mu_2)/\sigma\} \\ &\equiv \log([x-\mu_1]-\mu_2)/\sigma, \end{aligned}$$

which implies both that

$$\begin{aligned} dG^{-1}F(x)/dx &\equiv d\{(\log[x-\mu_1]-\mu_2)/\sigma\}dx \\ &\equiv \{\sigma[(x-\mu_1]\}^{-1} \end{aligned}$$

and

$$\frac{1}{dG^{-1}F(x)/dx} \equiv R(x) \equiv \sigma[x-\mu_1],$$

since $R(x)$ is the reciprocal of the derivative of $G^{-1}F(x)$.

Hence, for any logmodel (such as the lognormal) the derivative of R is a constant function of x, specifically $R'(x) \equiv \sigma$, and thus the second derivative of $R(x)$ *at any point* is zero. Any logmodel is completely characterized by the three parameters: μ_1, μ_2 and a parameter σ numerically equal to the property $R'(x_0)$, where x_0 takes on any real value. The three parameters μ_1, μ_2 and σ illustrate just how careful one must be when parameters are identified with properties. That the same symbol, σ, was used in the representation of the cdf model $G[(x-\mu)/\sigma]$ and the cdf logmodel $G\{(\log[x-\mu_1]-\mu_2)/\sigma\}$ does not necessarily mean that this symbol corresponds to the same properties of both curves. Specifically, while the parameter σ of the model $G[(x-\mu)/\sigma]$ is almost universally associated with the property standard deviation, for the logmodel the parameter $1/\sigma$ of $G\{(\log[x-\mu_1]-\mu_2)/\sigma\}$ is more closely associated with a property skewness than it is with the property standard deviation. Specifically, if $1/\sigma$ within $G\{(\log[x-\mu_1]-\mu_2)/\sigma\}$ increases while μ_1 and μ_2 are held constant, the skewness of the curve thus represented increases. Alternatively, if $1/\sigma$ and μ_2 are held constant and μ_1 decreases, then skewness increases. We will see below that this last connection between skewness and the value assumed by the μ_1 parameter of $G\{(\log[x-\mu_1]-\mu_2)/\sigma\}$ has many practical implications.

Unlike previous definitions of skewness and kurtosis, definitions based on $R(x)$ can be closely associated with distribution parameters. Just as the $x/G^{-1}F(x)$ scale-function introduced in the previous chapter is constant for any model, $R'(x_0)$ is constant for any argument x_0 for any logmodel $G\{(\log[x-\mu_1]-\mu_2)/\sigma\}$. A zero $R'(x)$ and constant scale

$x/G^{-1}F(x)$ of any model corresponds to zero $R''(x)$ and constant $R'(x)$ of any logmodel. Hence, the three curves, $x/G^{-1}F(x)$, $R(x)$ and $R'(x)$ can serve as the curve counterparts of the three properties, scale, skewness, and kurtosis, in the sense that the estimation of these three curves facilitates the selection from among three possible choices of representation (1) models, (2) logmodels, and (3) extended models, which differ from the expression $G\{(\log[x-\mu_1]-\mu_2)/\sigma\}$.

The estimation of all three members of the curve sequence $x/G^{-1}F(x)$, $R(x)$ and $R'(x)$ does not depend upon conventional likelihood or other procedures for the estimation of parameter values, but instead uses curve estimation techniques such as those described in Chapters 3 and 4. Hence, use of these curves bypasses the circular reasoning by which the choices made in the estimation phase of the statistical process depend upon the specification of the elementary function G, while the normal probability plot, lognormal plot and most other graphical or goodness-of-fit procedures for the selection of G depend on pilot parameter estimators.

In contrast to lambda and other approaches considered in previous chapters, the estimation of $R(x)$ and $R'(x)$ is accomplished by using what will be called below modular estimators, where two distinct multiplier sequences are used to determine two modular estimators from which, in turn, a single $R(x)$ curve is constructed, and where three distinct modular estimators are used to determine a single $R'(x)$ curve. In general, a modular estimator is to a sample what an extended model is to some population. Both are wholes for which it is of some advantage to distinguish between component parts. In particular, different multiplier sequences can be used to estimate different modules. In this way the whole can be optimized with respect to individualized component parts for both statistical and computational advantage.

7.2 The specifics of model, logmodel and fractional power model selection

Approaches based on R and R' are conceptually quite different from the process commonly called model fitting. They involve the parsing of a symbolic representation. In the context of written exposition, model fitting is equivalent to using a dictionary to find the best word. On the other hand, selection of a strategy for dealing with an extended model is like the study of a writing style. It is a parsing process where the best point of attack is located, and then the complex structure of an extended model (the analog of a prose sentence or paragraph) is systematically disentangled. $R(x)$ is one curve designed to parse the transformation v and model G, as well as the μ_1 and other parameter components of an extended model.

Below the cdf of an iid sample's elements, $F(x)$, will be assumed to be identically equal to the three-parameter cdf $G\{(\text{v}[x-\mu_1]-\mu_2)/\sigma\}$, where g, the derivative of G, will be specified explicitly as an elementary function, and v will either be the logarithm or some fractional power function $\text{v}(z)\equiv z^{[1/(1+K)]}$, for $K \neq -1$. The $R(x)$ function constructed from these curves is closely related to the curve $G^{-1}F$ estimated by quantile-quantile Q-Q plots (Wilk and Gnanadesikan, 1967). Gilchrist and Jewell (1972) described a procedure for using multiple Q-Q plots that relies on the parallelism of estimated curves, each individually associated with a user-selected constant, to indicate goodness of fit. According to Gilchrist and Jewell:

> Two extensions to . . . multiple Q-Q plots may be very useful. The first is to adjust the definition of the Q-Q plot in order to make the plot a horizontal line if the shapes of the distributions being compared are the same. It may be easier for the eye to check whether several lines are parallel if they lie on the horizontal.

For the standard normal special case, $G\equiv\Phi$, where $G'\equiv g\equiv\phi$, Tarter and Kowalski (1972) proposed and illustrated a technique that utilized a special case of $R(x)$, $\phi\Phi^{-1}F(x)/f(x)$, which is identical to the reciprocal of the derivative of $\Phi^{-1}F(x)$. For this and other $R(x)$ curves, L'Hopital's rule can be used to show that at any given x, $R(x)$ is the limit as $\epsilon\rightarrow 0$ of the scale parameter σ_x of the model $G[(x-\mu]/\sigma_x)$ which fits the v-model $G\{(\text{v}[x-\mu_1]-\mu_2)/\sigma\}$ perfectly at the endpoints of an interval of length 2ϵ centered at x. This simple relationship had been discussed only in the special lognormal case by Cohen (1951), who indicated that the ". . . standard deviation of the individual observations is approximately proportional to the magnitude of the observations." As $\epsilon\rightarrow 0$, $R(x)$ is the limit of the scale parameter σ_x of the model $G[(x-\mu]/\sigma_x)$, which fits the v-model $G\{(\text{v}[x-\mu_1]-\mu_2)/\sigma\}$. Thus, evaluations of this curve *are* what Cohen calls a "standard deviation." Moreover, $R(x)$'s linearity is common to all logmodels, $G\{(\log[x-\mu_1]-\mu_2)/\sigma\}$, and does not depend on G being a normal cdf.

For $G\{(\text{v}[x-\mu_1]-\mu_2)/\sigma\}$, where v is any one-to-one function with a continuous derivative v$'$, $R(x)\equiv[\text{v}'(x-\mu)]^{-1}$. This relationship indicates why $R(x)$ is useful as a parsing tool for the logmodel special case of the v-model, and why powers of $R(x)$ can be effectively used in conjunction with v-models, which are associated with fractional power transformations. $R(x)$'s simplicity for the log and fractional power special cases of v stems from the simplicity of the reciprocals of the derivatives of these functions.

Consider any variate X that can be transformed to a variate $Y=(X-\mu_1)^{[1/(1+K)]}$, where $K \neq -1$ and where Y's cdf can be represented by model $G([y-\mu]/\sigma)$. The curve $R(x)$ estimated from an iid sample of these X variates has the property that

$$[R(x)]^{[(1+K)/K]} \equiv \sigma(X-\mu_1)$$

That $[R(x)]^{[(1+K)/K]} \equiv \sigma(X-\mu_1)$ implies $R(x) \equiv \{\sigma(X-\mu_1)\}^{[K/(1+K)]}$. When

$$\begin{aligned} R(x) &\equiv \{\sigma(X-\mu_1)\}^{[K/(1+K)]}, \\ R'(x) &\equiv \{K\sigma/(1+K)\}\{\sigma(X-\mu_1)\}^{[\{K/(1+K)\}-1]}, \end{aligned}$$

which implies that at any point x where $R'(x)$ is nonzero,

$$R(x)/R'(x) \equiv [(1+K)/K](X-\mu_1).$$

To prove that for any one-to-one cdf G,

$$R(x)/R'(x) \equiv [(1+K)/K](X-\mu_1)$$

is a sufficient condition for v to be a fractional power transformation $Y=[X-\mu_1]^{[(1/(1+K)]}$, note that $R(x)/R'(x) \equiv [(1+K)/K](X-\mu_1)$ is a first order differential equation whose general solution is

$$Ce^{\frac{K}{1+K}\int \frac{1}{X-\mu_1}dx} = Ce^{\frac{K}{1+K}\ln(x-\mu_1)} = C(x-\mu_1)^{\frac{K}{1+K}}.$$

When $Q(x) \equiv G^{-1}F(x)$, the differential equation $dQ/dx = C^{-1}(x-\mu_1)^{[-K/(K+1)]}$ has the general solution

$$\begin{aligned} Q(x) &= C^{-1}\int (x-\mu_1)^{[-K/(K+1)]}dx + C' \\ &= C^{-1}(x-\mu_1)^{[1/(K+1)]}dx + C', \end{aligned}$$

which, since $Q(x) \equiv G^{-1}F(x)$ and G is a one-to-one function, shows that the linearity of $R(x)/R'(x)$ implies that v must be a fractional power transformation $(x-\mu_1)^{[1/(K+1)]}$. The same logic that underlies this theorem implies that the linearity of $R(x)$ is a sufficient condition for v to be a log transformation (Tarter and Kowalski, 1972).

The curve $R(x)/R'(x)$ is functionally independent of both μ_2 and σ. However, suppose the estimated curve $R(x)/R'(x)$ is applied to determine estimators of both K and μ_1, specifically, $\hat{K}$ and $\hat{\mu}_1$. Then further suppose that by means of the transformation $Y=[X-\hat{\mu}_1]^{\overline{1+K}}$ a sample of data is generated. Then the two parameters μ_2 and σ can be estimated using methods based on the model $G([y-\mu]/\sigma)$.

The curve $R(x)/R'(x)$ is not only linear for any fractional power v-model, but is linear for any logmodel. Specifically, $R(x)/R'(x) \equiv x-\mu_1$ for any logmodel. Note that while the line $y=x-\mu_1$ has a slope of one, the slope of $y=[(1+K)/K](X-\mu_1)$ appoaches one as $K \rightarrow \pm\infty$. This relationship indicates that any logmodel can be treated as some limiting case of any fractional power v-model where $K \rightarrow \pm\infty$ within $Y=[X-\mu_1]^{[1/(1+K)]}$. Thus, the estimated slope of $R(x)/R'(x)$ can be used graphically to distinguish between the assumption of some finite K fractional power transformation and the assumption of some logmodel's validity, in much the same way that the slope of $R(x)$ itself can be used to distinguish between logmodel $G\{(v[x-\mu_1]-\mu_2)/\sigma\}$ and model $G[(x-\mu_1)/\sigma]$.

7.3 Extended model and general curve estimator complexity

$R(x)$ and $R(x)/R'(x)$ are two examples of curves by which the fit of a model like the normal or extreme value, can be compared to the fit of an extended model like the lognormal or three-parameter Weibull, or representation where v is a square root transformation and G is the normal or extreme value model. Generally speaking, a two-parameter model's goodness of fit is implied by the linearity of a horizontal $R(x)$ line, while the goodness of a three-parameter logmodel's fit is implied by the linearity of a *slanted* line. In practice when $R'(x)$ is evaluated at or near an estimated mode, this estimate of skewness discloses how important it is to use a three-parameter logmodel generalization of a two-parameter model; for example, a three-parameter general Weibull rather than a two-parameter extreme value model. Note, since the three-parameter Weibull extended model is to the extreme value model what the three-parameter lognormal is to the normal model, both the lognormal and this Weibull are special cases of the v-model $G\{(v[x-\mu_1]-\mu_2)/\sigma\}$.

The determination of extended model complexity by means of $R(x)/R'(x)$ corresponds to the application of a stopping rule to select the appropriate number of terms of a series nonparametric estimator. Here, like the choice of an orthogonal system such as $\{\exp(2\pi ikx)\}$, the choice of a model such as G is an opening gambit that will direct the statistical game into certain channels. Ideally, however, as the game progresses, the relative contribution attributable to opening gambit G should decrease until, at the end of the spectrum, a result is achieved that could have been achieved by almost any choice of G. The curves $R(x)/R'(x)$ and $R(x)$ can help to determine where the best representation is located within the complexity spectrum; for example, whether K of a fractional power exponent should be treated as a parameter, and four-parameter closed-form expression used, whether a three-parameter logmodel should be selected, or whether a two-parameter model should be assumed to be an appropriate form of representation.

Although use of $R(x)/R'(x)$ followed by $R(x)$ resembles stopping rule selection of truncation point m of a Wahba raw series estimator, determination of degree of complexity for parametric and nonparametric expressions differ. The transition from model to logmodel to fractional power v-model representation is all or nothing, in the sense that no analog of a multiplier sequence is used in the way that several of the multiplier sequences discussed in previous chapters modify the KT crown of series-kernel representation. Nevertheless, use of an estimator of $R(x)/R'(x)$, followed by use of an estimator of $R(x)$, determines where the model, logmodel, fractional power v-model brim should meet the crown, in other words, the most appropriate degree of representational complexity. In this way these estimated graphs bring the flexibility of using the complex multiplier systems described in Chapter 4 to bear on choices within the model, logmodel, fractional power v-model spectrum.

ic7.4 Specific elementary function choices of G

It is useful to consider those aspects of the $G\{(v[x-\mu_1]-\mu_2)/\sigma\}$ parsing process that specifically pertain to the parameter μ_1. Because, unlike the two parameters μ_2 and σ, the parameter μ_1 is part of the argument of the function v, it plays a role which is considerably different than that played by either of the two other parameters of $G\{(v[x-\mu_1]-\mu_2)/\sigma\}$.

One limitation of conventional extended model $G\{(v[x-\mu_1]-\mu_2)/\sigma\}$ representation and estimation is the need for a pilot estimator of the pretransformation location parameter μ_1. In particular, when the logarithm or some fractional power is used as the transformation v extended model component, the traditional approach requires that the constant μ_1 be subtracted from each of the n variate values, $\{X_j\}$, $j=1,\ldots,n$. Not only the estimation of the parameters μ_2 and σ, but even nonparametric curve estimators computed from lognormal data can be very sensitive to the selection of the proper value of the constant $\mu=\tau$. While μ_1 is sometimes called a *threshold parameter* (Crow and Shimizu, 1977, p.3), in this chapter the term pretransformation location parameter was used because μ_1 satisfies the conditions necessary for a parameter to be a location parameter. For example, if a positive constant c is added to the x within $G\{(v[x-\mu_1]-\mu_2)/\sigma\}$ to form $G\{(v[x-(\mu_1-c)]-\mu_2)/\sigma\}$, this operation has the effect of translating the resulting curve c units to the left along the x-axis.

Note that there is no necessary reason why some function v cannot be used within some form of v-model representation $G\{(v[x-\mu_1]-\mu_2)/\sigma\}$, where v is defined for both negative and positive arguments. It happens, however, that since v is commonly applied as a transformation that "corrects" for the skewness of curves which have support on a ray which begins at some finite point and extends to plus infinity, functions like the log, fractional power, arcsin, and other common choices of v, are undefined for negative arguments. Since this is the case, in later chapters we will drop the distinction between components of v's argument as location parameters μ_1 and as threshold parameters τ, and use the more commonly applied symbol τ. It will also be understood that v is only defined for values $x>\tau$. Once again, the ambiguity of closed-form representation illustrated by v's argument either serving as a location or a threshold parameter, provides an example of the care that must be taken when individual components of a closed form expression are associated with individual properties of a curve.

Once the exact value of the threshold parameter has been subtracted from certain variates, the natural log of these differences is often assumed to have a normal, exponential, extreme value, logistic, or Cauchy $G([x-\mu_2]/\sigma)$. Because all these distributions have been studied extensively, it is much more convenient to estimate the post-transformation location parameter μ_2 than the pretransformation location or threshold parameter μ_1. Once data have been transformed to either the normal, or to some other standard, and therefore extensively studied model, the estimation of the post-transformation location parameter μ_2 is reduced to the problem of estimating the single location parameter of a normal or other much studied variate.

Methods whose detailed implementation was described in Chapters 3 and 4 were used to estimate the three curves shown in Figures 7.1. These curves are based on estimates of the particular lognormal density with parameters $\tau=10$, $\mu_2=0$ and $\sigma=1$ chosen for this experiment. The curve shown in Figure 7.1A is an estimated pdf obtained from an iid sample $\{Y_i\}$ of 1000 of these lognormally distributed variates. The curve shown in Figure 7.1B was obtained by applying the same curve estimation methods used to obtain the dotted curve based on data consisting of the 1000 transformed variates $\{Y'=\log(Y_i-10)\}$. To demonstrate the effect produced by estimating τ to be a value close to, but not exactly equal to its true value, the same curve estimation techniques were used a third time, based on the variates $\{Y'=\log(Y_i-9)\}$. This resulted in the characteristically highly asymmetric curve shown by the dashed curve in Figure 7.1C.

A normal probability plot can be used to ascertain the value of the parameter μ_2 when $\mu_1=\tau$ is known. Because it is necessary to repeat trials with a variety of user selected guesses of τ's value, the parameter τ is computationaly difficult to estimate. Hence, in most applications it is preferable to determine the values of μ_1 and σ graphically, rather than the values of μ_2 and σ.

As will be discussed below, the computation of $R(x)$ for the normal model differs from certain other examples of $R(x)$ computation. Specifically, the following two-step algorithm describes how the $R(x)$ curve corresponding to four logmodels can be estimated:

Step (1), a nonparametric estimator of the cumulative $y=F(x)$ is substituted

into $1-y$, for the power-exponential family,

into $-y\log(y)$ for the Weibull-extreme value family,

into $y(1-y)$ for the loglogistic-logistic family, and

into $\{1+[\tan\pi(y-1/2)]^2\}^{-1}/\pi$ for the family containing the logmodel such that G is the standard Cauchy cdf, which below will be called the logCauchy.

Step (2), the function computed in Step (1) is divided by an estimator of the pdf $f(x)$. Here estimator modularity can take advantage of the use of separate procedures to determine the separate f and F components

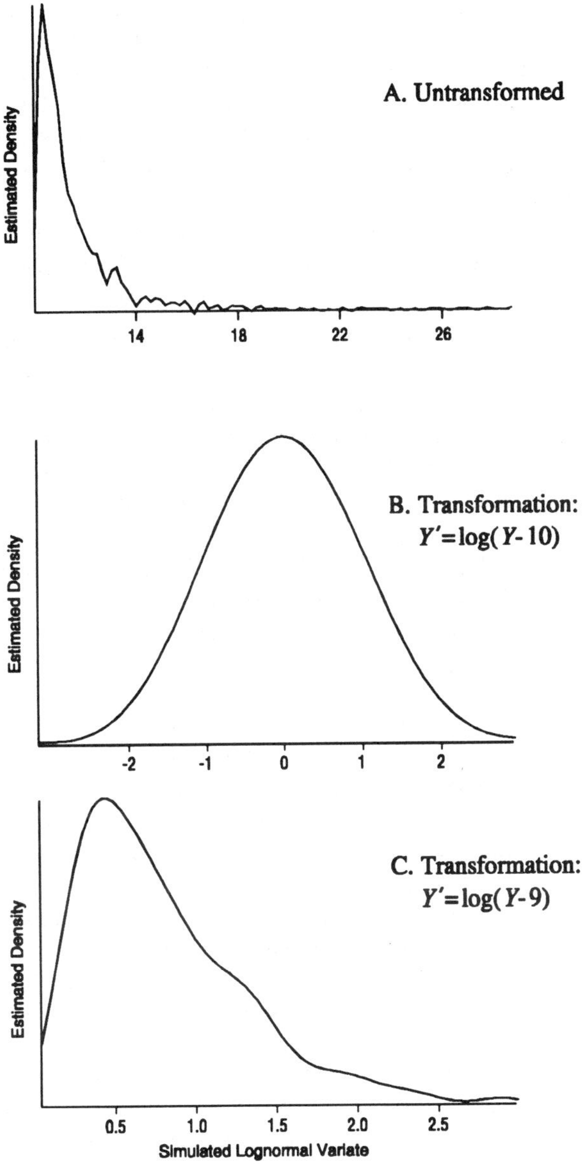

Figure 7.1. *Density estimates constructed from a sample of 1000 points from a lognormal distribution with* $\mu_1=10$, $\mu_2=0$, *and* $\sigma=1$. *Two different transformations were used for the estimates labeled B and C.*

of $R(x)$ based on individualized multiplier systems. The use of two multiplier systems permits a latitude of choice that can be applied to take advantage of the known role played by each estimator module within the composite estimator. For example, the $R(x)$ composite curve is a fraction with f as its denominator. Since division by a small number has the effect of amplifying inaccuracies in this number, it is best either to graph $R(x)$ only for estimated values of f that exceed a given value or, alternatively, to choose an estimator of f that makes use of procedures described in Chapter 4 for tail region accuracy enhancement.

For the four special cases of $R(x)$ defined above:

(A1) Whenever the true distribution is exponential, extreme-value, logistic, or Cauchy; $R(x)$ will be a horizontal line of height equal to the value assumed by the scale parameter of the model.

(B1) Whenever the true distribution is power, three-parameter Weibull, loglogistic, or logCauchy; $R(x)$ will be a diagonal line whose slope and y intercept are determined by the parameter σ and pretransformation location parameter $\mu_1=\tau$.

(C1) The $R(x)$ curves will be functionally independent of the post-transformation location parameter μ_2. Yet they can be used to determine the difficult to estimate parameter $\mu_1=\tau$.

There are also two general relationships which pertain to all $R(x)$ curves:

(A2) As $\mu_1=C$ increases, logmodel $G\{v[x]\}$, where $v[z]=\{C[\log(z+C)-\log C]-\mu\}/\sigma$, approaches the model $G[(x-\mu)/\sigma]$. This limit relationship explains why a logmodel $R(x)$ with any slope $\epsilon>0$ will, as $\epsilon\rightarrow 0$, approach the $R(x)$ curve of a model. In general, any model, like the normal or extreme value closed form expression, is the limiting case of some logmodel, like the lognormal or three-parameter Weibull.

(B2) The $R(x)$ of any of the above logmodels is the limiting case as $K\rightarrow\infty$ of the $R(x)$ of the v-model where v is the fractional power $v(z)=z^{[1/(1+K)]}$. Hence, that $[R(x)]^{[(1+K)/K]}$ is linear for some finite K, implies that some fractional power transformation is a more appropriate choice of v than is the log function. However, the larger this finite K is, the less difference it makes whether this v or the log transform is selected.

Note that the graphing of $R(x)$ does not require prior knowledge of any parameters, as do most other model or transformation selection methods. This facilitates $R(x)$'s role in the parsing process of prying v-model components apart.

Some further mathematical derivation is required to establish the validity of (A2) and (B2) because both of these relationships depend on the limit of a function of a function. Limit relationships, such as the limit of the quotient of two functions, often are valid only under special conditions. Taylor (1955) shows that the limit of a sum (or product) is equal to the sum (or product) of the limits and the limit of a quotient is equal to the quotient of the limits, provided the limit of the denominator is not zero. Because power, Weibull, loglogistic, or logCauchy $R(x)$'s are elementary functions, these theorems are all that are needed to establish (A2) and (B2) for these models. However, the normal cumulative Φ is not an elementary function. Thus, the properties of limits outlined by Taylor do not necessarily apply to a limit of the normal cumulative whose argument is some nonidentity second function.

Theorem: As τ approaches minus infinity, any normal, power, Weibull, loglogistic, or logCauchy logmodel cumulative $G\{(v[x-\mu_1]-\mu_2)/\sigma^*\}$ with pretransformation location parameter $\mu_1=\tau$, post-transformation location parameter equal to $[\log|\tau|+\mu/\sigma]$ and scale parameter $\sigma^*=\sigma/|\tau|$, approaches the corresponding model cumulative $G[(x-\mu)/\sigma]$.

Proof for the normal special case: Suppose a cumulative can be expressed as $F(\eta(x))$ where $\eta(x)\equiv\{[\log(x+C)-\log C]/C^{-1}-\mu\}/\sigma$, for $x>C$. Consider the series expansion for $F(x)$: $\Phi(x)=(1/2)+(2\pi)^{-1/2)}\{x-(x^3/6)[1-(3/10)(x^2/2!)+\ldots]\}$. Each bracketed term is less than or equal to a consecutive term of the power series expansion of $\cos(x)$. By Abel's theorem the power series expansion of $\cos(x)$ converges uniformly for any value of x. Thus, by the comparison test, the power series expansion of $\Phi(x)$ is uniformly convergent everywhere.

Let $\Phi_m(x)$ represent the m-term partial sum of the above power series. Consider the double limit

$$Lim_{C\to\infty}\, Lim_{m\to\infty}\, \Phi_m\left\{\frac{\log(x+C)-[\log C+\mu/C]}{(\sigma/C)}\right\}.$$

Each individual $\Phi_m(x)$ is a finite sum of powers of the curly bracketed expression within the double limit. Since $\Phi_m(x)$ was shown above to approach Φ uniformly, Theorem 7.11 of Rudin (1953) applies, and the order of the two limits of the double sum can be reversed. L'Hopital's rule implies that $Lim_{C\to\infty}[\log(x+C)-\log C]/C^{-1}=x$, which proves the theorem.

7.5 $R(x)$ and modularity

The construction of the curves $R(x)$ and $R(x)/R'(x)$ differs from the construction of other curve estimators considered in previous chapters. That two distinct component curves are used to construct $R(x)$ and three curves are needed to construct $R(x)/R'(x)$ both complicates certain procedures and offers the opportunity to enhance estimator efficiency by making use of the substantial literature which deals with generalized estimators of the curves f and F.

Besides the advantage of modular latitude of choice, discussed in the previous section, by utilizing $R(x)$ and $R(x)/R'(x)$ we can deal with model and transformation specification simultaneously by using closely related geometric curve characteristics. Like many other graphical tools, $R(x)$ is designed to accomplish a data analysis task based on the premise that the human eye is an excellent detector of the slant and straightness of a line, but is not particularly sensitive to other curve characteristics. This is the reason why it is desirable for $R(x)$ to estimate a horizontal line in the situation where a model provides adequate representation of a density curve, and a slanted straight line when a logmodel provides better curve representation than does the corresponding model. The effective use of $R(x)$ depends on nonparametric curve estimators to break down a complex configuration (embodying several underlying assumptions) into its component parts. Some parts, like G and v, are themselves functions, while other parts, like μ_1, μ_2 and σ are parameters. This allows a statistician to parse a symbolic representation of an extended model in much the same way that a grammarian parses a sentence.

The particular modular approach based on $R(x)$ brings the large body of literature devoted to the estimation of basic curves, such as densities and cumulatives, to bear on problems formulated in terms of an important class of extended model—the v-model. Yet all these problems can be dealt with by taking advantage of the one curve that the human eye can discern with great precision, the straight line.

For example, the linearity of powers of $R(x)$ or the linearity of $R(x)/R'(x)$ can be used to select the constant K of the fractional power transformation $v(z)=z^{[1/(1+K)]}$ (Freeman and Tukey, 1949, 1950). Here the $K=-2$ case corresponds to the reciprocal transformation, while $K=1$ corresponds to the square root transformation. That $R(x)$ is linear, but not a constant function, implies that some log function is an appropriate choice for v. That $R(x)$ is a constant function implies that the model-limiting case of some logmodel is the appropriate choice of cdf. The choice of whether to use some trial power of $R(x)$ or, alternatively, $R(x)/R'(x)$ to estimate K directly, depends upon the efficiency of the modular estimator components. However, both approaches have the following advantage over most alternative goodness of fit procedures:

It is usually difficult to distinguish systems of curves such as $F_1(x)=G_1\{(v_1[x-\mu_{1,1}]-\mu_{2,1})/\sigma\}$ and $F_2(x)=G_2\{(v_2[x-\mu_{1,2}]-\mu_{2,2})/\sigma\}$ that involve multiple elementary functions. Comparing the fit of $F_1(x)$ and $F_2(x)$ by matching $F_1(x)$ and $F_2(x)$ to the sample cumulative requires estimation of six parameters: $\mu_{1,1}$, $\mu_{2,1}$, σ_1, $\mu_{1,2}$, $\mu_{2,2}$ and σ_2. The modular $R(x)$ and $R(x)/R'(x)$ methods are based only on the estimation of pdf and cdf modules and consequently do not require that any model or extended model parameters be estimated. Since $R(x)/R'(x)$ involves the derivative of $R(x)$ the construction of this curve is based on the estimation of the three curves F, f, and f', the derivative of f discussed in Section 3.2, and thus, like the use of $R(x)$, does not require that likelihood or other methods be used to determine parameter estimators.

A major advantage of the modular approach is that the assembly of composites of these three curves can be studied and applied from a broad base of information, in terms of the curve estimators described in Chapters 3 and 4 as well as many other alternative forms of nonparametric curve estimators. The literature of closed form parametric model fitting is full of special and pathological cases where, for example, maximum likelihood estimation procedures yield equivocal results. Consequently, the comparison of models where likelihood or other estimators are substituted for parameters seems in many instances to be of questionable value, since accurately estimated parameters within a poor fitting model may yield results that are superior to those obtained from poor estimates of the parameters within a correct choice of model. Naturally, it is also true that not all curve estimators will estimate all curves well all of the time. But it seems preferable to deal with the peculiarities of a few very general curves, such as the triad, F, f and f', rather than the peculiarities of the dozens of models and extended models which, even in the univariate continuous case, fill compendia such as Volumes 1 and 2 of Johnson and Kotz (1970).

7.6 $R(y|x)$ curves

Like the bivariate and multivariate generalizations of the curve $f_\lambda(x)$ described in Section 6.6, $R(x)$ and $R(x)/R'(x)$ curves can easily be generalized to conduct studies based on bivariate data. We will discuss only one such generalization, based on the connection between R and the parameter σ, here, and will consider more complex applications in Chapter 9, which deals with bivariate survival estimation and other bivariate applications.

The data used to construct the curves shown in Figure 7.2 were obtained from the California Department of Health Services, CDHS, and consisted of soil lead measurements in parts per million (PPM) for front yards of a sample of residences in Oakland, CA. Blood lead data

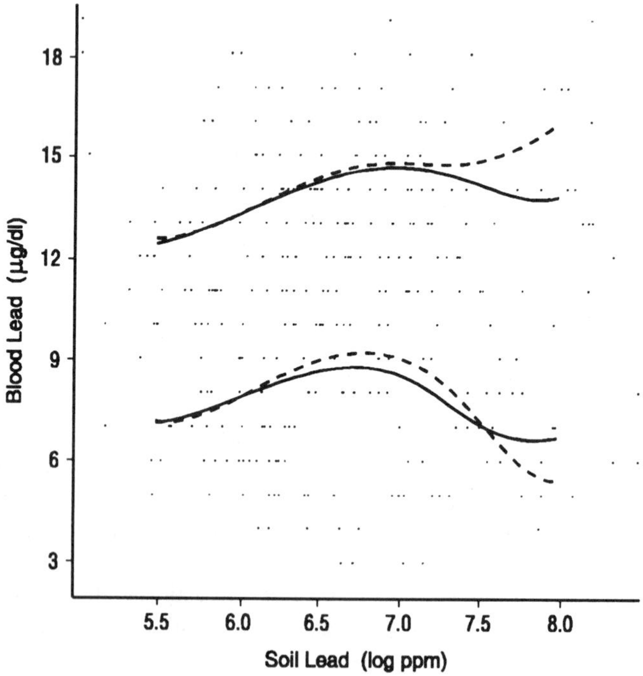

Figure 7.2. *The solid curves form a conditional box-plot based on an estimate of the regression curve $\sigma(Y|x)$; the dashed box-plot is based on nonparametric estimates of the .25 and .75 quantiles. The data are the amount of lead measured in front yard soil and the blood lead level of 337 children.*

contained samples taken from 337 children ages 1–5 who live in these residences. Blood was taken by capillary finger tip or by venipuncture and was measured in micrograms/deciliter ($\mu g/dl$). CDHS analyzed all samples of blood and soil for lead using atomic absorption spectrophotometry (Kizer et al., 1989).

To illustrate how $R(y|x)$ curve estimators can be applied, Figure 7.2 displays two sets of bands and the data points used to compute both sets of bands. Each point's x and y coordinates were obtained by measurement of log soil lead exposure attributable to the front yard of a child's house, and the levels of lead present in that child's blood, respectively. Both the band-pair based on the estimates of $\sigma(Y|x)$, which are plotted as a connected curve, and the bands plotted with dashes, are two dimensional generalizations of classic box-plots (Tukey 1977, p.40). While box-plots are commonly used to display estimated quantiles of a density $f(y|x)$, when x is a *qualitatively* valued variate, the bands shown in Figure 7.2 are different, in the sense that here x of $f(y|x)$ is a running value assumed by the quantitative *continuous* variate, soil lead. A detailed description of the computational and statistical procedures by which this form of $f(y|x)$ can be constructed are detailed in the next chapter.

The upper and lower edges of a box plot are constructed to enclose 50% of total estimated probability mass or, more specifically, estimate the .25 and .75 conditional univariate distribution quantiles. Two distinctly different procedures are used to do this. Specifically, for $j=1,2,\ldots,50$, the points $y_{x_j}^{(.25)}$ and $y_{x_j}^{(.75)}$ determine the curve shown by the dashes, where

$$\int_{-\infty}^{y_{x_j}^{(.25)}} \hat{f}(z|x_j)dz \equiv \hat{F}(y_{x_j}^{(.25)}|x_j) \equiv 1-\hat{F}(y_{x_j}^{(.75)}|x_j) = .25.$$

Although it resembles the above empirical conditional box plot, the bands drawn as solid curves in Figure 7.2 used R-based methods where the G component of R was chosen to be the standard normal cdf. For each index $j=1,2,..,50$ of an equally spaced sequence of log soil x coordinates, a conditional pdf $f(y|x_j)$ and cdf $F(y|x_j)$ were estimated. Then the sequence of curves $R(y|x_j)$, $j=1,2,\ldots,50$, was estimated and, finally, for each j a local estimator of the parameter $\sigma_{y|x}$ was obtained by substituting the estimated mode $\hat{\mu}_j$ corresponding to the highest estimated value of $f(y|x_j)$ for the y argument of $R(y|x_j)$. As described in

Tarter and Freeman (1987) this procedure will yield the curve $\sigma(Y|x)$ when data have any bivariate normal density. In this case each $R(y|x_j)$ curve will be a horizontal line whose height, $\sigma_{y|x}$, provides a local estimator of the conditional density scale parameter. Since for the normal cdf, Φ, $\Phi(-.674)=.25$ and $\Phi(.674)=.75$, the edges of the estimated conditional box plot were calculated using the relationship $\hat{\mu}_j-.674R(\hat{\mu}_j|x_j)$, $j=1,2,\ldots,50$, for the lower edge and $\hat{\mu}_j+.674R(\hat{\mu}_j|x_j)$, $j=1,2,\ldots,50$, for the upper edge.

The bifurcation of the two sets of bands at the right hand side of Figure 7.2 underscores three differences between regression curves based on $R(y|x)$ and many other forms of nonparametric and parametric regression curves which target the distribution properties $E(Y|x)$ at a sequence of x values. (1) The curve $R(y|x)$ is an intermediary, or means to an end, focused on the scale parameter $\sigma(y|x)$ and not on some location parameter μ_x, the parameter usually associated with the properties $E(Y|x)$. (2) $R(y|x)$ is based on techniques described in this chapter which target model and extended model parameters by means of the basic properties of curve estimators of densities, f, and cdfs, F, as distinguished from properties such as $E(Y|x)$.

(3) At any given value of x_j, $R(y|x)$ provides an estimate of the x-specific scale parameter regression curve, $\sigma(y|x)$, by information that is local to $R(\hat{\mu}_j|x_j)$. In other words, it is only within some small neighborhood of $\hat{\mu}_j$ that the constituent curve estimators of f and F components of R affect the regression curve. The estimated conditional mode $\hat{\mu}_j$ is, of course, usually far from the tails of the distribution. Consequently, the usual sensitivity of estimators of such curves as $E(Y|x)$ and $\sigma(y|x)$ to outliers or, in for case being discussed, to the dual component mixture structure of the Oakland soil-blood lead distribution, is greatly reduced.

In Chapter 9 we will return to bivariate applications of R and R' curves. It will be shown that besides linear relationships involving the shape of these curves themselves, useful linear relationships can be obtained by using $R(y|x)$ and $R'(y|x)$ to estimate the curves $\tau(x)$ and $K(x)$ which regress the threshold parameter and fractional power exponent against the value x assumed by some independent variable. For example, when $K(x)$, estimated using $R(y|x)$ and $R'(y|x)$ curves based on the cdf G, is not a horizontal line, this implies that a single fractional power transformation of the dependent variate Y to a new variate Z with generating function G does not exist. When $\tau(x)$, estimated using $R(y|x)$ and $R'(y|x)$ curves based on the cdf G, is not a horizontal line, this implies that even if a single fractional power transformation of the dependent variate Y to a new variate Z with generating function G does exist, there is no single threshold parameter τ within the fractional power transformation $v(Y)\equiv(Y-\tau)^{[1/(1+K)]}$ that can be used to create a new variate Z with generating function G. In other words, the fractional power transformation that creates the new variate Z must be of the form $v(Y)\equiv[Y-\tau(x)]^{[1/(1+K)]}$, where the nonhorizontal $\tau(x)$ modifies the variate to account for the relationship between the threshold at which a variate relationship begins, and the value x of the independent variate in question.

7.7 Truncation adjustments

The curve $\tau(x)$ in some ways resembles, and in some ways differs, from the curve $\rho(x)$ discussed at the beginning of this chapter. When $\rho(x)$ is zero before some value x_0, following which $\rho(x)$ is nonzero for all $x>x_0$, then in the following sense the relationship between the Y variate and the x variable could be said to "begin at the x_0 level of x." Were variate dependence to be wholly determined by $\rho(x)$ and were x_1 and x_2 both less than x_0, then knowledge that differentiates x_1 from x_2 would not differentiate the conditional density $f(y|x_1)$ from the conditional density $f(y|x_2)$. However, this would not be true if x_1 and x_2 were both greater than x_0.

Unlike the assertion that the *relationship* between Y and x begins at x_0, when either or $\log[Y-\tau(x)]$ or $[Y-\tau(x)]^{[1/(1+K)]}$ for $K>0$, is found to be the correct choice of data transformation, the *support of the conditional density of* Y given x begins at $\tau(x)$. In other words, for these transformations to be valid, at a given value of $x=x_0$ no measurement of Y can be smaller than $\tau(x_0)$: hence, the appropriateness of the term threshold, which connotes that there is an absolute lower limit to the value assumed by the Y variate which, when $\tau(x)$ is a non-horizontal line, is determined by the value assumed by the variable x. Like $\rho(x)$, $\tau(x)$ describes an aspect of the dependence of Y on the value of x. However, this aspect is of a highly specialized form, specifically, that at a given x it is impossible for Y to be less than $\tau(x)$.

The role the parameter τ plays in the expression $\log[Y-\tau]$ differs considerably from the role it plays in the expression $[Y-\tau]^{[1/(1+K)]}$. This difference raises certain complex problems involving the issue of density truncation. For example, for $K>0$, the smallest value assumed by $[Y-\tau]^{[1/(1+K)]}$ is always taken in practical applications to be zero; while

$\log[Y-\tau]\rightarrow-\infty$ as $Y\rightarrow\tau$ from above. Yet, as suggested at the end of Section 7.2, and as discussed in more detail below, as K increases it matters less and less whether an appropriately scaled logmodel, or the fractional power v-model, $v(z)=[z]^{[1/(1+K)]}$, is chosen as an argument of G. But, for finite K, logmodels differ from fractional power v-models, in the sense that the $v(z)$ argument of $G\{[v(z)-\mu_2]/\sigma\}$ assumes a different minimum value for these two forms of representation.

The above reasoning implies that for a cdf $G\{[v(z)-\mu_2]/\sigma\}$, where $v(z)=[z]^{[1/(1+K)]}$, the region over which the argument of G is defined has $-\mu_2/\sigma$ as its left endpoint. Specifically, if the cdf $F(z)$ is defined over the entire real axis, then $G\{[v(z)-\mu_2]/\sigma\}$ should be multiplied by the indicator function $I_{(C,\infty)}(z)$, where $C=-\mu_2/\sigma$. But, with the notable exception of the exponential model, most commonly used choices of G have support on the entire real axis. This suggests that many fractional power $G\{[v(z)-\mu_2]/\sigma\}$ models can be classed as truncated forms of representation, in the sense described by Kendall and Buckland (1960) who indicate "that a truncated distribution is formed from another distribution by cutting off and ignoring the part lying to the right or left of a fixed variate-value." Here, for example, the other distribution can be extreme-value, logistic, Cauchy, or normal.

Specifically, if g represents the standard form density that corresponds to one of the above cdfs G, and if C is a point of truncation which satisfies the inequality $\tau\leq C$, then the left truncated fractional power density model with support on the interval $C<x$ equals

$$\frac{(x-\tau)^{-\frac{K}{K+1}}}{(K+1)\sigma}\left\{\frac{g\left[\left((x-\tau)^{\frac{1}{(K+1)}}-\mu_2\right)\Big/\sigma\right]}{1-G\left[\left((C-\tau)^{\frac{1}{(K+1)}}-\mu_2\right)\Big/\sigma\right]}\right\}I_{(C,\infty)}(x).$$

For example, for a G that is the standard normal cdf and when $C=\tau=\mu_2=0$ and $K=\sigma=1$, the left truncated fractional power model is a chi-square density with one degree of freedom.

By taking the power series expansion of e^z, where $z=[\log(x-\tau)]/[K+1]$ it can be shown that as $K\rightarrow\infty$ any left truncated fractional power v-model approaches a left truncated logmodel with scale parameter σ^* and post-transformation location parameter μ^* if

$(K+1)\sigma \rightarrow \sigma^*$ and $(K+1)(1+\mu_2) \mapsto \mu^*$. Note that this mathematical relationship can be combined with the Theorem (A2) of Section 7.4 to show that any left truncated model based on standard form cdf G is a limiting case of the above truncated fractional power v-model.

Cohen (1988) surveys ML estimation procedures for estimating the parameters of left, right, and both left and right (doubly), truncated lognormal special case v-models. Among the many references listed at the end of Cohen's chapter there is no mention of any literature that deals with the lognormal as a special case of the larger family of logmodels, or even fractional power v-models with G taken to be the normal cdf Φ. Thus, very little is known about alternative methods that, like R-based methodology, can be used to compare the goodness of fit of truncated fractional power v-models, logmodels, and models based on a wide variety of standard cdfs G. As we will see in the next two chapters, exploratory curve estimation techniques described in this chapter can help answer questions regarding the representation of conditional densities of some Y variate given one or more variable values; in other words, goodness of fit, for both untruncated and truncated elementary functions in $p>1$ dimensions.

There is an approach that can often be used to adapt the methodology described in this chapter to deal with truncated forms of what would, in the untruncated case, be elementary functions. The procedure illustrated below, or some other method can be used to estimate the degree to which the curves f and F are truncated, where the untruncated cdf G model and its derivative g is used to compute $R(x)$. Then if the cut off and ignored part lying to the right, left, or both the left and the right, of a fixed variate-value has total probability mass P, it follows that both the pdf and cdf curve estimator components of $R(x)$ should be multiplied by $(1-P)$. If the $R(x)$ curve is evaluated over a region which is substantially separated from the region over which parts of f and F are truncated, then multiplication of estimators of both f and F by $(1-P)$ has the effect of nullifying the distortion produced on the unmodified $R(x)$ curve attributable to truncation. In Section 9.3 we will see how P can be estimated in bivariate survival analysis applications.

In certain univariate applications there is a simple procedure that can be used to estimate the parameter P of right and left truncated densities. For example, the pdf f's estimator, shown as a solid line in Figure 7.3, was obtained by applying the estimated Watson multipliers discussed in Section 4.6 to the 940 data points remaining after $n=1000$ N(0,1)

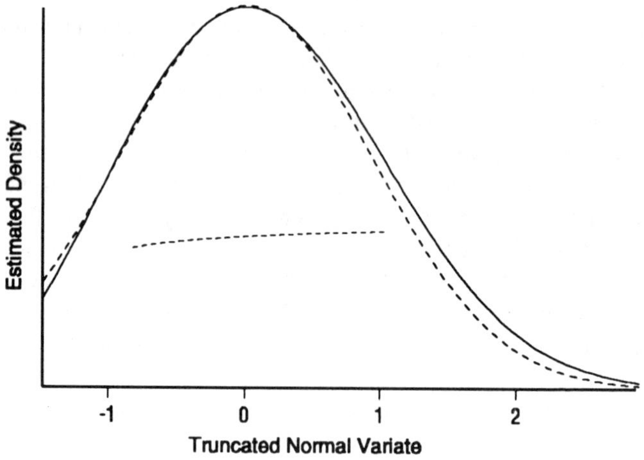

Figure 7.3. *The solid curve is an estimate of a normal density truncated at* $x = -1.5$*; the dashed curve is a fitted normal density.*

variates were left-truncated by removing all points whose value was less than -1.5. The dashed curve was a fitted normal obtained by setting the location parameter equal to the estimated mode X(MODE) and using an estimator of $R(X(\text{MODE}))$ to determine the fitted normal's scale parameter. Here, the estimated R curve is shown below the empirically and parametrically determined curves.

Although the method described below applies whenever a unimodal distribution's mode equals a model's location parameter, we will illustrate it here in the normal case. If the density f is truncated normal such that $P<0.5$, then the height of f at the mode, $f(X(\text{MODE}))$, equals the standard untruncated normal density evaluated at zero divided by the product of the scale parameter of the untruncated normal and $(1-P)$. Therefore, under the null hypothesis of normality, the relationship

$$\Phi\left[(1-P)\left\{\frac{X(MIN)-X(MODE)}{\sigma^*}\right\}\right]=P$$

applies, where σ^* is the height of the standard normal density at its mode, 0.3989, divided by $f(X(\text{MODE}))$. Here X(MIN) designates the smallest point within the support region of the left truncated density. By using the sample minimum minus a slight adjustment which our computational routine automatically uses to avoid plotting problems, here X(MIN) was estimated to be -1.548. Again, using Watson multipliers, $f(X(\text{MODE}))$ was estimated to be 0.4255.

Evaluation of the pdf estimator shown as a solid line in Figure 7.3 yielded 0.9379 as an estimate of σ^*. (If σ^* had simply been estimated as the height of the estimated R curve at the estimated mode, then a value 0.9333 would have been determined as the approximate value of σ^*. That there is very little difference between 0.9376 and 0.9333, as well as the high degree of horizontal linearity of the estimated R curve shown in Figure 7.3, illustrates the high degree of robustness R commonly exhibits in the presence of small degrees of truncation.)

To illustrate how robust R actually is, note that for the 0.9333 value of σ^*, and $P=.06$, the argument of the Φ function would be estimated to be -1.5670. Using the Pearson and Hartley (1956) tables, and linear interpolation of table values, the standard normal cdf Φ evaluated at -1.5670 equals 0.05856, which is very close to the true value, $P=.06$. Based on the 0.9376 value of σ^* the argument of the Φ function was estimated to be -1.5598. The standard normal cdf Φ evaluated at -1.5598 equals 0.05938. Of course, it should be noted that in practice the equation for P must be solved iteratively by using σ^*, estimated as the height of the estimated R curve at the estimated mode, where R is based on the previous estimate of P.

As this simple example illustrates, based on the density estimation procedures described in Chapters 3 and 4, accurate estimates of small to moderate degrees of truncation P can be obtained from samples of the size similar to the size of the sample discussed here. As far as alternative approaches are concerned, consider the likelihood function of a size n uniformly distributed sample with density $\mathrm{f(x)}=\mathrm{I}_{[0,1]}[(x-\mu)/\sigma]/\sigma$, where I is the previously defined indicator function of the unit interval. Obviously, any alternative choices of parameter values within the likelihood function L, other than the values of μ and σ which yield an f which extends from the smallest to the largest sample elements, will reduce the value of L

According to Gumbel (1958, p. 60) no matter what density f is, the probability that the largest of n past iid observations with density f will never be exceeded in n future similar trials is equal to 1/2. This illustrates that the likelihood approach, used in conjunction with the above f, or alternatively, with some elementary function multiplied by this f, tends to yield highly biased estimators, since in even as few as n additional trials the chance is 50-50 that an observation will occur outside of the support of f estimated using likelihood procedures. Of course, as shown by Rosenblatt (1956), all absolutely continuous density estimators must be biased to some degree, and consequently, it is yet to be proven whether curve estimators evaluated at their modes or standard likelihood procedures provide better ways to deal with truncated models.

CHAPTER 8

Applications of likelihood multipliers

Once the best fitting model G and transformation v of the expression $G\{(v[x-\mu_1]-\mu_2)/\sigma\}$ is selected, the following problem still remains: How can the parameter μ_2 be estimated? Maximum likelihood (ML) estimation is the most common applied statistical approach used for this, and for many other purposes. Consequently, one application described in this chapter involves the ML estimation of location parameters by specialized multiplier sequences.

There are two reasons why ML estimation using multipliers differs from other likelihood procedures. The first is, that like the box-plots displayed in Figure 7.2, specialized regression curves can be designed. Application of these curves can utilize simple, and yet rapid, algorithms for the computation of Fourier coefficients of a conditional density in terms of the Fourier coefficients of a joint density. The second difference from other forms of likelihood stems from the modularity through which data, in the form of the trigonometric moment equally spaced evaluations of the sample characteristic function, are multiplied term by term by likelihood multipliers. This modularity makes it convenient to perform data correction operations on the sample trigonometric moments rather than on the data points themselves.

Figure 8.1 illustrates how likelihood multipliers can be used to perform a task with nonparametrically obtained regression curves, which closely resembles a useful modification of classical forms of parametric regression. Shown are those members of a sample of $n=1000$ iid data points with common density $f(x,y)$ that occur within a region extending in both the x and the y direction from -2.5 to 2.5. To illustrate just why likelihood multipliers are useful, four regression curves are displayed after being determined from the $n=1000$ sized sample of

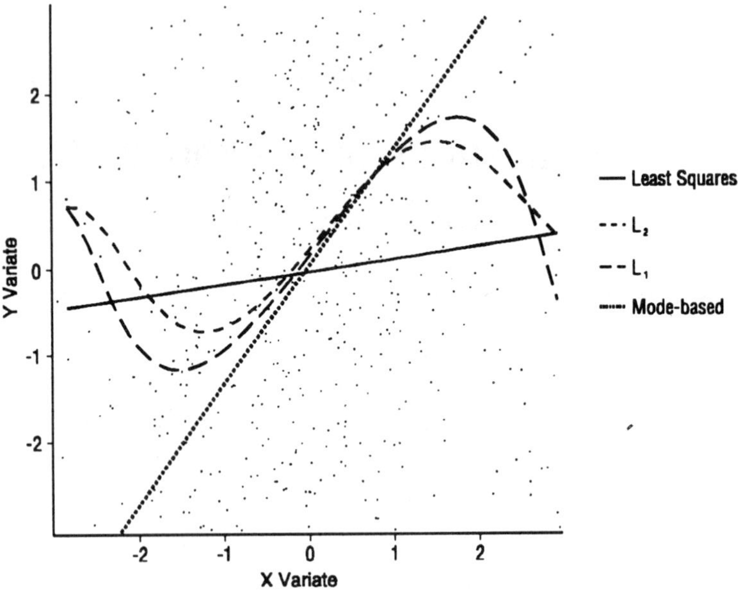

Figure 8.1. *Four different regression curves superimposed on a mixture of 50% correlated bivariate normal data and 50% uniformly distributed points.*

$$f(x,y) = (\tfrac{1}{2})\mathrm{N}(x,y|\mu_x = \mu_y = 0, \sigma_x = \sigma_y = 1, \rho = 0.5) + (\tfrac{1}{2})\mathrm{I}_{[-3.5,3.5]}(x)\mathrm{I}_{[-3.5,3.5]}(y)$$

data, in other words, from a mixture consisting of bivariate normal variates with zero means, unit standard deviations, correlation one-half and mixing parameter one half, mixed with uniformly distributed background fog variates.

The solid and straight curve is a conventional least-squares (LS) fitted line, while the seemingly linear curve formed from small squares was obtained by plotting the estimated modes of a sequence of conditional density estimators using procedures described in the previous section. Both the curve formed of short dashes and the curve formed of long dashes were obtained by using likelihood multipliers of the sort described in later sections of this chapter. These curves differ from each other in only one respect: the short dashed curve was constructed by using a set of likelihood multipliers associated with a normal model $g=\phi$, where, of course, the normal is the model for which the parametric likelihood algorithm, and simple least-squares equations, yield identical curves. The long dashed curve was constructed by using a set of multipliers associated with the Laplace model rather than the normal model corresponding to the dotted curve. The Laplace model has the sample median as its ML estimator. Consequently, the long dashed curve can be characterized as being a form of median-likelihood regression while the short dashed curve can be characterized as being a form of mean-likelihood regression.

It is shown in Tarter (1979a) that despite the intrinsic truncation which is fundamental to Fourier series expansion over a finite interval, likelihood estimators based on Fourier techniques are identical to their parametric counterparts for the sample mean and sample median, and many other commonly used statistics, with the proviso that data points are transformed within the unit interval leaving sufficient margins between data extremes and the unit interval boundaries. For example, formulas for the sample absolute deviation and standard deviation in terms of the sample trigonometric moments, which will be described in Chapter 10, are only valid if data are transformed within the interval [1/4,3/4] (Brunk, 1977).

One advantage to using a Fourier formulation of basic likelihood relationships is that it permits regression curves to conform to a wide variety of shapes, while at the same time it retains many of the properties of conventional forms of regression. For example, the long dashed curve is closer to the conventional LS line, because, like its parametric counterpart, it is strongly drawn toward outlier fog points. On the other hand, like those forms of parametric regression discussed later in this chapter, which are based on the Laplace model and least total residual absolute values (LAV), rather than total squared deviations, the median likelihood regression curve based on LAV likelihood multipliers are outlier resistant.

Before turning to the details of likelihood multiplier applications, it is relevant to mention that in keeping with the approaches described in the last chapter, likelihood multiplier algorithms require only minimal adaptation to be used with a large number of representational choices. Conventional parametric likelihood computation can involve very different formulas and algorithms for even slightly different models. On the other hand, with the one exception that different multiplier systems were selected, the same computational routine was used to construct both the short and long dashed curves in Figure 8.1. Thus, the computational modifications required to obtain the short and long dashed curves shown in this figure differ solely in same way that routines described in Chapter 3 for cdf estimation differ from some basic pdf multiplier-based curve estimator. In fact, the construction of these two regression curves, any of the pdf and cdf curve estimators described in Chapter 3, and even the mixture component variance reduced density estimator described in Chapter 5, all depend on the same fundamental algorithm.

8.1 Likelihood multiplier LAV and other types of regression

Before describing the LAV-type semiparametric approach, it is useful to compare five different regression curves that, like the LAV-type procedure, are alternatives to a parametric representation of some curve $\beta(x)$ fit to data using the least-squares criterion. The first alternative is the mode-based regression introduced in Chapter 1 and depicted in Figure 1.9 as well as the curve drawn as a sequence of small squares shown in Figure 8.1, where, rather than targeting the expectation $E(Y|x)$, regression is based on the population mode or modes $\text{Mode}(Y|x)$.

To illustrate this procedure using real, rather than simulated, data, the estimated modes $\hat{\mu}_j$, $j=1,2,\ldots,50$, where the jth mode corresponds to the highest estimated value of $f(y|x_j)$, are graphed in Figure 8.2. These modes were estimated from the sample of 337 children ages 1–5 who lived in Oakland, California residences, where the Y-variate, blood lead, and the X-variate, soil lead measured in each home's front yard, were described at the end of the last chapter. However, below and to the right of the curve formed from the modes at which each $f(y|x_j)$, $j=1,2,\ldots,50$, takes on its absolute maximum value, is a second short curve. This curve corresponds to a sequence of the values at which those relative modes $\hat{\mu}_j^{(r)} \neq \hat{\mu}_j$ occur, where $f(\hat{\mu}_j^{(r)}|x_j) \leq f(\hat{\mu}_j|x_j)$ such that $f(\hat{\mu}_j^{(r)}|x_j) \geq f(y|x_j)$ for all y within some neighborhood of $\hat{\mu}_j^{(r)}$, $j=1,2,\ldots,50$.

As described by Tarter (1979a, expressions (2.22) and (2.23)), if $B_{s,v}$ represents a bivariate Fourier coefficient, then the conditional density of y given x_0 can be represented as

$$\sum_{v=-\infty}^{\infty} B_v^{(x_0)} e^{2\pi i v y}, \quad \text{where } B_v^{(x_0)} = \frac{1}{f(x_0)}\left[\sum_{s=-\infty}^{\infty} B_{s,v} e^{2\pi i s x_0}\right],$$

and f is the marginal density of the x variate. There are many ways to use this expression to perform the computations that underlie regression

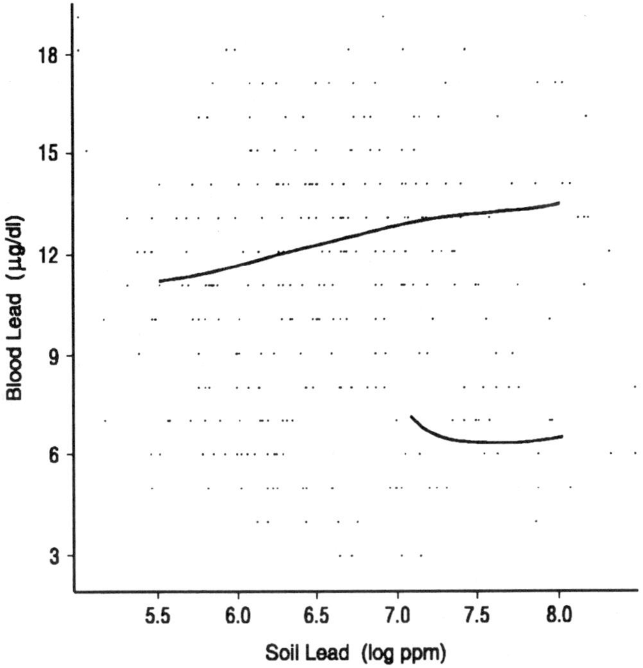

Figure 8.2. *Mode-based regression indicating two relationships between the amount of lead measured in front yard soil and the blood lead level of 337 children. A marginal log transformation has been applied to the soil lead variate.*

curve construction. However, all applications share a common feature whose advantages and disadvantages can be described by a comparison with bivariate statistical methods that involve allocation of data points among contiguous strips. These methods were commonly used many years prior to the development of modern forms of regression.

There was no discussion of nonparametric regression in theoretical statistics text-books before 1963, when Fisz introduced this subject. However, the literature of applied statistics does describe a curve fitting procedure, the method of averages (MA) based on allocation of data points among contiguous strips. With minor modifications the MA can easily be applied for purposes similar to that served by types of nonparametric methodology now available. How it can be thus applied is informative in terms of the distinction between nonparametric and semiparametric forms of regression.

Lipka's book, *Graphical and Mechanical Computation* (1918), presented the MA and, the conventional least-squares (LS) algorithm, where the MA, like LS, is described as a *model-fitting* procedure, and not as a *nonparametric* estimator. To use the MA, bivariate data are first allocated among contiguous strips with borders parallel to the dependent variate's y-axis so that each strip contains approximately the same number of data points. Individual independent and dependent variable means are then computed from the points within each strip.

The implementation of the MA described by Lipka did not use the locus of consecutive mean pairs to form a model-free regression curve, or fit a segmented polynomial regression model, as described by Gallant and Fuller (1973). Instead, either two such mean pairs, each substituted into the equation for a line, provided the system of simultaneous equations needed to fit a two-parameter model or, to deal with more complex forms of relationship for $q>1$, $q+1$ such mean pairs from $q+1$ contiguous strips were used to fit $q+1$ parameter equations, such as the polynomial model $E(Y|x)=\beta_0+\beta_1x+\beta_2x^2+\ldots+\beta_qx^q$.

It would be feasible to base either nonparametric or semiparametric regression on the MA by simply connecting consecutive strip mean pairs. Nevertheless, prior to 1963, elementary-function-based models provided the only form of representation applied for what were classed as "regression" purposes. As far as terminology is concerned, if a parametric procedure like conventional maximum likelihood (ML) were used to estimate the location parameter of the density of strip-specific ordinates by assuming a parametric model, this could be referred to as being a semiparametric procedure; parametric in terms of the strip-specific model, nonparametric in terms of relationships between individual strip-specific estimators. On the other hand, if a property estimator, like the sample mode of the strip-specific data points, were utilized, then the resulting curve could be said to be obtained nonparametrically, since no particular class of elementary function needed to be assumed.

Note, that the latter nonparametric procedure is not necessarily superior to the former semiparametric procedure in terms of generality. The only reason why the sample mean and median, and not the sample mode, are closely linked to particular models like the normal and Laplace elementary functions is that location parameter likelihood computations are particularly simple for these two models. The advantage of the mode, as discussed cogently by Scott (1992), is not that it is inherently more nonparametric than the mean or median, but rather that in many instances it provides more useful information about basic distributional structure than do the mean or median.

The distinction between regression based on properties, and regression based on parametric forms of representation, evolved surprisingly slowly. Fisz, 1963, authored a statistical text that distinguished elementary-function-based-parametric from model-free regression. Before 1963, techniques like the MA were used exclusively as regression model-fitting procedures, not as nonparametric or semiparametric procedures, in the sense of circumventing the need to represent the regression curve as some specific elementary function.

The curve commonly targeted by both the MA and LS methodologies, the polynomial model $E(Y|x)=\beta_0+\beta_1x+\beta_2x^2+\ldots+\beta_qx^q$, provides a good vantage point from which to compare generalized and conventional forms of regression. Consider the assertion that: "With sufficient data the value of q in the polygonal model can be increased to efficiently obtain a satisfactory fit to almost all forms of data likely to be encountered in realistic situations." To illustrate one instance where this statement is misleading, consider a modification of the MA where the observed mode or, more importantly, observed *multiple* modes of the estimated conditional $f(y|x)$, of the strip-specific dependent variate values, are used to form a nonparametric regression curve. Were there two population subgroups with appropriately selected location and mixing parameters, any estimate of $E(Y|x)=\beta_0+\beta_1x+\beta_2x^2+\ldots+\beta_qx^q$ would be less informative than a mode-based, Mode($Y|x$), nonparametric curve, even when Mode($Y|x$) is estimated by the MA based on a few broad strips.

The above example underscores a basic limitation of completely parametric forms of regression methodology. No matter what size q is used, a polygonal regression curve must be a single-valued function. Acceptance of this limitation is surprising in light of the substantial interest taken by Karl Pearson and other pioneer statisticians in the problem of distribution superposition and cluster analysis, which even in the bivariate case can involve dual variate relationships. From the hindsight of today's immense nonparametric regression literature, contributed to by Watson (1964), Nadaraya (1965), Brunk (1965), and later many others, it is unclear why regression analyses are so restrictively concerned with single-valued relationships while cluster and decomposition problems are seldom formulated in terms of conditional models. By this point in the history of statistics, Klonias (1984), Yakowitz and Szidarovszky (1985), as well as many others, have illustrated not only the feasibility, but the practical utility of nonparametric regression methodology.

A complete exposition of the field of nonparametric and semiparametric regression is out of the question in this short monograph. Consequently, we will only consider those aspects of this field that make use of the computational and other advantages of series-based methods. We will also only discuss those nonparametric likelihood methods that apply the multiplier-based techniques introduced in Chapters 2 and 3. Wegman (1970a,b) and Silverman (1978b) include excellent discussions of general maximum likelihood estimation theory and applications, while Tapia and Thompson (1978); Good and Gaskins (1970, 1980) and Silverman (1982) describe the details of univariate penalized likelihood procedures.

Just as there is a close correspondence between the series methods described in Chapters 2 and 3 and kernel methodology, with only two exceptions there is an almost perfect compatibility between those forms of series-based penalized likelihood functions recommended by Good and Gaskins (1980), and the univariate methods described in this chapter. One exception is that the decomposition technique described in Chapter 5 induces a negative degree of smoothness that acts counter to the tendency of penalized likelihood procedures to smooth estimators.

Suppose a Fourier series based on the orthogonal system $\{\exp(2\pi ikx)\}$, obtained by penalized likelihood or by any other method, has coefficients that, when divided by the sample trigonometric moments, yield a sequence of complex numbers that have real and imaginary terms, both of whose absolute values are less that one. Since, in practice, any such series must contain only a finite number of terms, these complex numbers can be treated as multipliers, in the sense that when used as Fourier coefficients, they will uniquely determine some kernel.

Even if the above process does not yield a sequence of complex numbers that have real and imaginary terms whose absolute values are less that one, the coefficients of penalized likelihood expansions can still be used within most of the likelihood estimation procedures described in this and later chapters. However, for the following reason, there is very little point to following this line of inquiry further now. This is because hardly anything is known today about bivariate and higher dimension penalized likelihood. On the other hand, as suggested by the close association between series based on $\{\exp(2\pi ikx)\}$ and the kernel and shifted histogram, (Scott, 1992), approaches, which predominate within today's multivariate curve estimation and regression literatures, there are very simple ways that bivariate and multivariate series can be estimated using the MISE-based techniques described in Chapters 3 and 4.

There are two practical advantages to regression methodology based on bivariate Fourier expansions whose s,vth term will be represented below as $B_{s,v}$. (1) These expansions can be applied in a modular fashion with the mixture decomposition techniques described in Chapter 5 as well as the model selection procedures described in the previous two chapters, which can help determine which likelihood multiplier system is the most appropriate for a given data set. (2) This methodology avoids a commonly encountered data analysis dilemma that plagues processes for histogram and clinical life table construction as well as procedures based on allocation of data among contiguous strips.

Strip, histogram class interval and life table construction all involve data fragmentation that leads to the following decision process: by making strip width very small, the bias intrinsic to using a single $E(Y|x)$ to represent all x values within the strip can be reduced. However, by making strip width narrow, and the portion of the sample within the strip small, the variance of any estimator of $E(Y|x)$ will be increased. This trade-off between bias and variance or estimator inefficiency is sometimes characterized as a trade-off between validity and reliability (Bechtoldt, 1958, pp. 1245-1247).

When methodology based on $B_{s,v}$ is used, the entire sample of n measurements contributes to the estimation of $B_v^{(x_0)}$ at any given x_0. Consequently, the basic trade-off inherent to the user selection of strip width is replaced by the design of the bivariate multiplier evaluations of some bivariate KT. While, generally speaking, conditionals cannot be estimated quite as precisely as can marginals, in the main, the bias-efficiency dilemma can be circumvented. The bias-efficiency dilemma is characteristic of methods that operate in data space. The use of frequency domain implementation can bypass this dilemma by harnessing the intrinsic smoothness of Fourier series estimators.

In addition to regression curves based on the mode, Mode($Y|x$), as discussed by Scott (1992); Silverman (1981), Tarter (1981), and Tarter, Cooper and Freeman (1983), there are many other possible useful types of multiplier-based nonparametric regression. Certain of these procedures can deal with a data analysis problem that already has a substantial *parametric* literature. It may be desirable to increase resolution or enhance contrast to better understand individual population subgroups. There are also many situations where members of specific subgroups constitute "outliers," while some other subgroup may be of primary scientific importance. For these reasons, the robustness of an estimator with respect to outliers is much sought after.

With respect to outliers, nonparametric and semiparametric approaches span two extremes. What we will call below, "LAV regression," is positioned at the spectrum end opposite to what can be characterized as "the Mode($Y|x$) regression" spectrum end. The LAV procedure combines the least-absolute-value fitting criterion with likelihood methodology. When there are two compactly distributed yet separated mixture components, rather than enhancing the resolution of an anomalous data structure, as does mode-based regression, LAV regression is insensitive to the presence of anomalies. Specifically, it enhances the resolution of the more dominant of two-component mixture model components.

The distinction between squaring a difference in conventional least-squares regression, and the role of squaring in MISE-based curve estimation, is central to the spectrum extremes referred to above. For MISE-based semiparametric methods the squaring operation is part of a metric which assesses goodness-of-fit. What is squared is not a distance between data points and the fitted curve, but a difference between an estimated density and its estimator. This latter difference stresses those portions of unimodal density support regions that are near the mode and, for most common distributions, yields estimates which are more precise than estimates over other regions.

Like the Mode($Y|x$) curves shown in Figure 8.2, the two curves shown in Figure 8.3 also exhibit the multicomponent structure of log soil lead, blood lead distribution. The regression curve depicted using long dashes was constructed by the following process.

Step (1): A sequence of conditional estimators of $\{f(y|x_j), j=1,\ldots,50\}$ was obtained by using the relationship between the univariate conditional Fourier coefficients, for example, if $x_j=x_0$, $B_v^{(x_0)}$, and the bivariate Fourier coefficients, $B_{s,v}$ for integer s and v.

Step (2): All members of $\{f(y|x_j), j=1,\ldots,50\}$, for example, $f(y|x_0)$, were modified using lambda multipliers to enhance the contrast between components so that there was an imperceptible degree of component overlap.

Step (3): In the case being illustrated, where the mixture appeared to be comprised of two distinct subpopulations, the densities of each subpopulation were estimated at each x_j by (3a) blanking out one of the two components and then, (3b) reversing the lambda contrast enhancement process using the same computational routine that executed Step

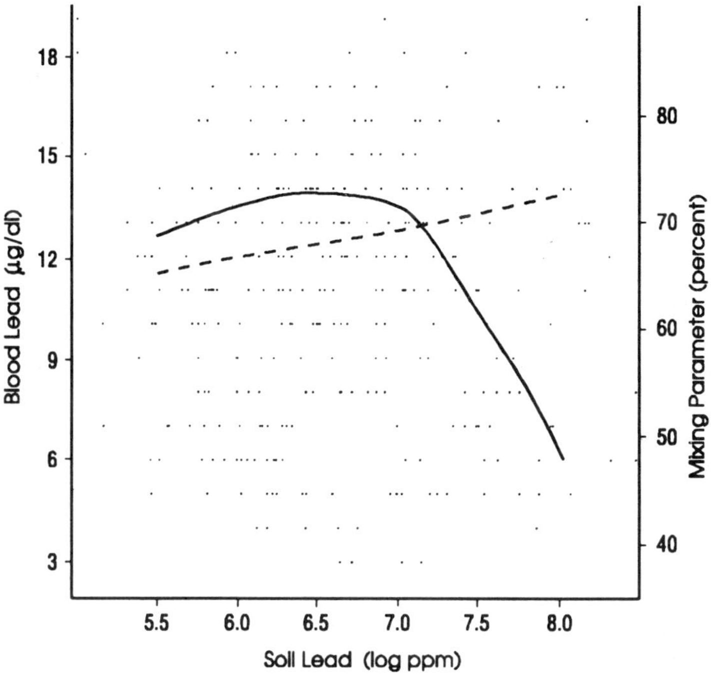

Figure 8.3. *The dashed curve is the estimated location parameter of the upper component of the same distribution estimated in Figure 8.2, i.e., the group with high blood-lead levels. The solid curve is a regression of the mixing parameter that illustrates how the proportion of the population in the upper component changes as the amount of soil lead changes.*

(2), but now with a choice of lamba which equaled minus the value used in Step (2), for example, $-\lambda_0$. This separation process was illustrated by Figures 1.7 and 1.8 and also discussed in Sections 5.4 and 5.5.

Step (4): The parameters of the component with the larger location parameter were estimated by the methodology described in the previous chapter,

Step (5): The larger of the estimated component y-variate location parameters $\mu_y^{(2)}(x_j)$, $j=1,2,\ldots,50$ was evaluated and graphed to obtain the dashed curve of Figure 8.3.

Below, the jth member of the sequence of lower location parameters will be represented as $\mu_y^{(1)}(x_j)$. Here the higher mixture component, $\mu_y^{(2)}(x_j)$, regression curve is nearly coincident with the upper mode-based regression curve shown in Figure 8.2. Generally, this will only be true when either: (1) the graphed mixture component is located at a considerable distance from other components or (2) as is the case here, all other components have very small mixing parameter values. Note that typically, when (1) and (2) both apply, data points from components with very small mixing parameter values are designated as outliers and dealt with using one of the specialized procedures reviewed by Barnett and Lewis (1984).

Regression based on the decomposition of a mixture of two bivariate normals has been considered by Quandt (1958, 1972), and Quandt and Ramsey (1978). In one sense, what Quandt and Ramsey call switching regressions are similar to the loci, $\mu_y^{(1)}(x_j)$, $\mu_y^{(2)}(x_j)$, $j=1,2,\ldots,50$. However, Quandt and Ramsey switching regression curves differ from these loci in a crucial sense. The assumption upon which their form of switching regression depends is that the mixture in question is a mixture of two bivariate normals. It follows from this assumption that the regression curves involved must be a pair of straight lines. On the other hand, the loci, $\mu_y^{(1)}(x_j)$, $\mu_y^{(2)}(x_j)$, $j=1,2,\ldots,50$ need not conform to any straight line, elementary function, or other model.

Tarter and Silvers (1975) applied a variant of the lambda procedure to the mixture considered by Quandt and Ramsey, which was based on the assumption that the mixture being decomposed was comprised of two bivariate normal components. Note, however, that while the conditionals of a bivariate normal must be univariate normals, the assertion that the conditionals of a bivariate distribution are univariate normals does not necessarily imply that this bivariate distribution is bivariate normal. It is this fact that makes it practical to devise a variant of the Quandt and Ramsey form of switching regression which is far more general than regression based on mixtures of bivariate normal components.

Suppose each member of the sequence $\{f(y|x_j)\}$ can be separated into two or more components. Then the parameters of each of these components can be estimated, and the curves obtained by graphing the estimated sequence of any one of these component parameters will not necessarily be lines, or any partially predetermined curves. This underscores the semiparametric nature of this form of regression, where the mixture model is assumed for each individual member of $\{f(y|x_j), j=1,2,\ldots,50\}$, but no assumption is made concerning the functional relationship between the parameters of sequence members $f(y|x_u)$ and $f(y|x_v)$ where $u \neq v$. Below, we will refer to this class of regression procedure as conditional switching regression.

The generalization of switching regression to *conditional* switching regression makes it possible to construct a useful semiparametric statistical tool that, at present, has no parametric counterpart. The mixing parameter p rather than the location parameter μ can be studied in terms of its relationship between two variates. The curve $E(Y|x)$ describes one aspect of the relationship between Y and one or more key or nuisance variates. Suppose, however, that the value assumed by a continuous response variate Y is principally of value as an indicator of membership in a diseased or hypersensitive mixture component, say the component whose parameters are labeled with the subscript 1. When $p(x)$ is estimated, it is not the change in the location of a subpopulation that is described, but instead, what is described is the chance that an as yet ungrouped observed measurement $X=x$ belongs to the group individualized by the subscript 1. The curve $p(x)$ can be highly informative, since it links the health/disease or normal/hypersensitive dichotomies to the value of x directly, and not indirectly, as do curves that estimate $E(Y|x)$.

For example, suppose that the problem of interest is whether or not some exposure to an environmental contaminant, such as soil lead, sensitizes some individuals to the contaminant. Sensitization can be defined as the functional relationship between the probability, $p(x)$, of membership in a sensitive group, and the exposure variate value, x. On the other hand, as shown by the solid curve of Figure 8.3, some protective mechanism or distinguishing characteristic may cause the curve $p(x)$ to decline as x increases. This $p(x)$ curve describes how the mixing parameter of the high blood lead distribution component changes as the log soil lead variate changes. Even more dramatically than both the mode-based form of regression shown in Figure 8.2, and the box-plot curves shown in Figure 7.2, the mixing parameter conditional switching regression curve is greatly affected by the anomalous high exposure, low response, subpopulation.

Figure 8.4 shows a conventional least-squares (LS) line enclosed by a set of conventional 95% confidence intervals constructed from the Oakland soil and blood lead data. Neither the confidence bands nor Fisher's z calculated from the sample Pearson correlation $r=0.071$ indicates a significant relationship between the variates. In comparison with the other forms of regression shown in Figures 8.1 and 8.2, the LS curve makes the worst of a bad bargain. Unlike the mode-based regression curve, it cannot be split into two components. Unlike the mixing parameter conditional switching regression curve it cannot indicate that

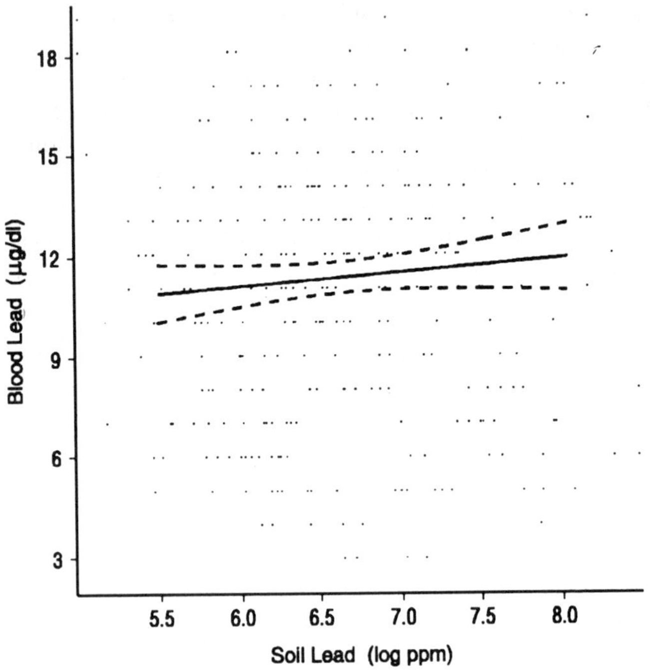

Figure 8.4. *Ordinary least squares regression line and 95% confidence bands superimposed on same data set described by Figure 8.2. Neither the bands nor Pearson's correlation were significant.*

subgroup membership is related to the log soil lead variate. However, even though it cannot disclose that subgroups composed of outliers exist, as does Mode($Y|x$) by bifurcation, or as does $p(x)$ by radically changing its curvature, the LS curve is highly sensitive to the presence of outliers. Thus, it cannot disclose or alert the researcher to the presence of the very distributional feature to which it is highly sensitive.

The bibliography of Barnett and Lewis (1984) contains 27 pages of references concerning the outlier problem, many of which deal with the bivariate or multivariate case. Several of these papers describe a particularly clever and commonly used approach to outlier forms of contamination, M-estimators. According to Barnett and Lewis (1984, p. 79), "M-estimators (maximum likelihood type estimators) . . . [achieve] . . . desirable properties of robustness . . ." through the use of a composite of least-squares and least-absolute-value error functions.

We will show in the next section that the maximum likelihood estimator (MLE) of any Pearson family member location parameter can be expressed in terms of the sample trigonometric moments $\{\hat{B}_k\}$, (Tarter, 1981). Using similar procedures, Tarter (1975a,c; 1979) investigated series-based location parameter MLEs of a parameters corresponding to the family of models $Ce^{-c_1|x|-c_2x^2}$, where $c_1 \geq 0$, $c_2 \geq 0$, and C is a constant which ensures that over its support region the model's definite integral equals one. Like the Pearson family, this system, which contains the normal as a special case, yields MLE that can be expressed in terms of estimates of the Fourier coefficients $\{\hat{B}_k\}$. However, unlike the Pearson family, the $Ce^{-c_1|x|-c_2x^2}$ family contains the Laplace density as a special case. Because of the Laplace's connection with the median, series relationships based on $Ce^{-c_1|x|-c_2x^2}$ can be applied to design nonparametric regression curves that are intermediates between curves based on the mean and curves based on the median.

The Laplace density is to the least-absolute-value portion of the M-estimator error function, what the normal is to the least-squares portion. Therefore, using simple linear relationships between the bivariate Fourier coefficients and the Fourier coefficients of members of $\{f(y|x_j)\}$, it is possible to construct a form of regression which, like the curves evaluated by $\{\mu_y^{(1)}(x_j)$ and $\mu_y^{(2)}(x_j)$, $j=1,2,\ldots,50\}$, is semiparametric, in the sense that the estimated curve need not conform to any prespecified model. In addition, by utilizing the density $g(z) = Ce^{-c_1|x|-c_2x^2}$ a form of regression can be designed that is based on the combination of least-squares and least-absolute-value fitting criteria, which in some ways can be conceptualized as a semiparametric M-estimator. In Figure 8.5 the upper curve (which below we will designate by the symbol LS) was based on the least-squares, c_2z^2, and the lower curve (which below we will designate by the symbol LAV) the least-absolute-value, $c_1|z|$, components of g.

The LAV curve is less affected by the anomalous high exposure/low response component than is the LS form of conditional regression. However, both curves are less informative than the mode-based and conditional switching mixing parameter regression curves shown in Figures 8.2 and 8.3. These curves are good substitutes for conventional regression procedures in applications where linear or any other pre-specified regression model appropriateness is uncertain. When targeted curves are actually linear, and sample size n exceeds 30, semiparametric LS regression will yield curves that are all but indistinguishable from conventional parametric least-squares curves over that part of the x-axis

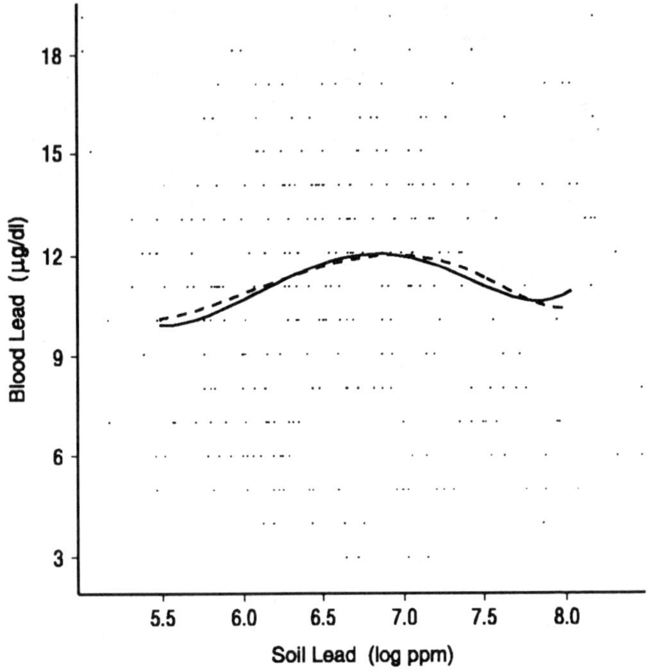

Figure 8.5. *Semiparametric regression curves. The dashed curve was based on least squares criteria while the solid curve relied on least absolute value. The data are the same as in Figure 8.2.*

equal to the X variate mean, plus or minus, two X variate standard deviations.

As shown in Figure 8.4, while even for large n the conventional least-squares curve responds to, yet cannot alert the researcher to, the presence of outliers, either the normal-LS, or Laplace-LAV, or any intermediate $Ce^{-c_1|x|-c_2x^2}$, $c_2>0$ and $c_1>0$, M-estimator conditional regression, will markedly curve toward the outlier subpopulation centroid. It is even possible to generalize $Ce^{-c_1|x|-c_2x^2}$ to be a mixture that is normal in a neighborhood of zero and Laplace elsewhere. Such a mixture can be used to construct a semiparametric procedure that corresponds to the most commonly applied form of parametric M-estimator.

Outliers and mixtures are only two out of many possible varieties of data anomaly. Consequently, after deriving certain basic relationships the next section, an application is described in Sections 8.4 and 8.5, which concerns the following practical aspect of likelihood implementation. Data are often gathered with a very limited degree of precision. For example, it is not at all uncommon for survival measurements obtained from patients suffering from a disease with a poor prognosis, like pancreatic cancer, to be specified as a whole number of months survived before death. Hence, for this, and many other response variates, only a very few data points assume a value greater than one of the twelve discrete whole numbered values, 0,1,...,11.

To be mathematically correct, the application of ML methodology based on any realistic iid sample requires the substitution of a discretized adaptation for some simple parametric model. For example, suppose $f(y|\theta)$were to represent the continuous density of actual survival times. If the variate value 1 were to designate that some individual survives at least one month, but dies before surviving two months, then the probability of 1's occurrence would equal $\int_1^2 f(x|\theta)dx$, not $f(1|\theta)$. This difference raises questions concerning the confounding of abridgment, binning or rounding choices with estimates of model parameters. Thus, there is a need for correction procedures that can be applied easily in conjunction with the methodology described in the next two sections, such as the extension to the frequency domain of a general correction approach developed by Craig and Sheppard (1936).

Much of the material reviewed in this chapter is based on the limit of the likelihood equation expressed as a series whose terms are products of the sample trigonometric moments $\{\hat{B}_k\}$ and the Fourier coefficients $\{\beta_k\}$ of the log of the assumed underlying elementary function model g. The next chapter will extend the application of this series to the generalized representation of the pdf. For either application, the sample trigonometric moments calculated from abridged, binned or grouped data can be first corrected for abridgment, or some other anomaly, and then substituted for $\{\hat{B}_k\}$ in the likelihood series. This general approach can be used with any assumed model g to circumvent substituting expressions like $\int_1^2 f(x|\theta)dx$ for expressions like $f(1|\theta)$ within the equations used to obtain likelihood estimators and to compute likelihood ratio tests.

To avoid confusion with the symbol μ, often used to represent the population mean, below we will use the letter θ rather than μ to represent the location parameter of the model $f(x)=g(x-\theta)$. We will also assume that $g(x-\theta)$ is some iid sample datum density such that g has support on the unit interval, over which $\log(g)$, the natural log of g, has a well defined Fourier transform. It will be shown that ML procedures for the estimation of θ can be expressed using Fourier series representation based on the complete sequence $\{\exp(2\pi ikx)\}$. The actual expressions for the log likelihood function, $\log L(\theta)$, that can be used for this application are, with one exception, identical to expressions used to form pdf estimators. The one exception is that information concerning the actual assumed standard model g determines a multiplier sequence formed from the Fourier coefficients of $\log(g)$. To distinguish this specialized sequence of multipliers from other forms of multipliers, below it will be designated by $\{\beta_k\}$ and this, and other general approaches to ML estimation in the frequency domain, will be characterized as forms of "trigonometric likelihood."

Representing $\log L(\theta)$ in terms of $\{\beta_k\}$ makes several procedures computationally convenient. A single algorithm can be used to calculate both (1) pdf estimators based on the sample trigonometric moments $\{\hat{B}_k\}$ and (2) $\log L(\theta)$. The $\{\hat{B}_k\}$ coefficients of these curves merely serve as estimators of the population trigonometric moments $\{B_k\}$. Hence, the question arises: For likelihood applications based on $\{\beta_k\}$, how can the most serviceable sequence of multipliers $\{b_k\}$ be determined so that $\{b_k\hat{B}_k\}$ best estimates $\{B_k\}$? To answer this and several other questions, a basic theorem will be presented in Section 8.4 that applies to all situations where an unweighted, $w(x)\equiv 1$, MISE criterion is used as a metric. This theorem concerns the connection between, on the one hand, the choice of $\{b_k\}$ and, on the other hand, the selection of multipliers like $\{\beta_k\}$.

Choosing $\{b_k\}$ independently of $\{\beta_k\}$ has many applications beyond and in conjunction with likelihood. For example, Chapter 5 described the application of lambda multipliers that can be used to estimate a curve $f^{(\lambda)}$, a mixture which differs from the density of sample elements, f, with respect to the second cumulants of its components. For this application, the Fourier coefficients of the density being estimated are not simply $\{B_k\}$ but are $\{b_k^{(\lambda)}B_k\}$, where $\{b_k^{(\lambda)}\}$ are the lambda multipliers defined in Section 5.4. Hence, the question arises: What multiplier sequence $\{b_k\}$ within $\{b_k b_k^{(\lambda)}\hat{B}_k\}$ provides the best MISE-based estimator of a curve with Fourier coefficients $\{b_k^{(\lambda)}B_k\}$? One corollary of the theorem discussed in Section 8.4 is that the choice of $\{b_k\}$ is functionally independent of the choice of user setting λ.

In summary, the same set of multipliers $\{b_k\}$ can be used iteratively to isolate an individual mixture component of $f^{(\lambda)}$ based on a sequence of user selected mixture second cumulant reducing choices λ within

$\{b_k b_k^{(\lambda)} \hat{B}_k\}$. Suppose the sequence $\{\hat{B}_k^{(\lambda)}\}$ contains this mixture component's estimated Fourier coefficients. The methods described in the previous chapter can then help determine which model g provides the best description of $f^{(\lambda)}$'s density. If $\{\beta_k\}$ represents the Fourier coefficients of $\log(g)$, we will now show that $\sum_{k=-\infty}^{\infty} k\hat{B}_k^{(\lambda)}\beta_k \, exp(2\pi ik\theta)=0$ will have a root equal to the trigonometric likelihood estimator of mixture component location parameter θ.

8.2 Series representation of the likelihood function

Suppose $\{f(y|\theta)\}$ represents a family of densities parametrized by θ, where θ is a scalar or a vector of parameter values, and the natural log of the density, $\ln[f(y|\theta)]$, has a uniformly convergent Fourier expansion over the unit interval [0,1] for all values of θ. For example, in Section 8.7 we will consider the case of any density which is a Pearson Family (PF) member that is truncated such that its support is on [0,1]. Applying integration by parts, a simple expression can be obtained for the maximum likelihood estimator of the location parameter of any PF density f. We will also assume that datum density f is non-zero and that $\log(f)$ is integrable within [0,1]. Were a datum density to be zero within any open subinterval M contained in [0,1], no data point at which the likelihood function is evaluated can occur in M. Hence, in those rare applications where it is necessary to apply these methods to a density h that is zero within some measurable interval M, h can be redefined as some function f identical to h over h's support region, but equal to some positive integrable function over M. Because these procedures only depend on the properties of f in arbitrarily small neighborhoods of the data points, points which cannot occur in M, this can be done without affecting any of the likelihood procedures described below.

Suppose $\{X_j\}$ is a sample of n iid variates with datum density f, where each X_j has been transformed within the unit interval [0,1]. As shown by Tarter (1979b), in terms of the m-term Dirichlet kernel, $K_D(x)=\sum_{k=-m}^{m} \exp(2\pi ikx)$, discussed in Chapter 3,

$$\int_0^1 \log[f(x|\theta)]K_D(x-X_j)dx = \sum_{k=-m}^{m} \int_0^1 \exp(-2\pi ikX_j)\log[f(x|\theta)]\exp(2\pi ikx)dx$$
$$= \sum_{k=-m}^{m} \exp(-2\pi ikX_j)\int_0^1 \log[f(x|\theta)]\exp(2\pi ikx)dx$$
$$= \sum_{k=-m}^{m} \beta_k(\theta)\exp(-2\pi ikX_j), \tag{8.2.1}$$

where

$$\beta_k(\theta) = \int_0^1 \log[f(x|\theta)]\exp(2\pi ikx)dx, \tag{8.2.2}$$

which in turn implies that

$$n^{-1}\sum_{j=1}^{n} \log[f(X_j|\theta)] = \mathrm{Lim}_{m\to\infty}\sum_{k=-m}^{m} \hat{B}_k\beta_k(\theta). \tag{8.2.3}$$

Since $\hat{B}_k$ is the mean of n bounded identically distributed variates X_j, $j=1,\ldots,n$, as $n\to\infty$ each $\hat{B}_k \to E_f(e^{-2\pi ikX_j}) = B_k$. Thus, (8.2.3) is a consistent estimator of the function of θ

$$\sum_{k=-m}^{m} B_k\beta_k(\theta). \tag{8.2.4}$$

If θ is a scalar location parameter, $\beta_k(\theta)=\beta_k\exp(2\pi ik\theta)$, where the sequence of multipliers $\{\beta_k\}$ are simply the Fourier coefficients of the logarithm of the density $f(x|0)=g(x)$. Thus, the methods outlined in the previous chapters can be used to estimate the likelihood limit (8.2.4). Also, as long as the series $(2\pi i)^s\sum_{k\neq 0} k^s\hat{B}_k\beta_k\exp(2\pi ik\theta)$ converges, this expression is the sth derivative of n^{-1} times the log likelihood function of the location parameter θ.

Trigonometric likelihood methods are quite different from ordinary likelihood methods based on elementary function models for

$f(x|0)=g(x)$. Estimator computation can be altered by simply substituting one sequence $\{\beta_k\}$ for another such sequence. This is a departure from conventional likelihood, which characteristically results in equations that differ substantially for different choices of g. In addition, the multiplier sequence $\{\beta_k\}$ can be combined with one or more other multiplier sequences to perform such useful tasks as grouping correction by using group corrected $\{\hat{B}_k\}$.

The utility of trigonometric location parameter likelihood stems, in part, from the modularity inherent in the use of the orthogonal system $\{\exp(2\pi ikx)\}$. Trigonometric likelihood modularity is based on the substitution of the estimated Fourier coefficients of some density different from the datum density f into expressions which involve the Fourier coefficients of f. For example, the following question is relevant when an estimated curve is based on the multipliers $\{b_k^{(\lambda)}\}$: When coefficients $\{b_k^{(\lambda)}\}$ of the lambda contrast-enhanced estimator are inserted as part of the multiplier system used to construct a curve, will the estimator optimality properties discussed in Chapters 3 and 4 apply to the modified estimator? We will now show that the optimal multiplier $\{b_k\}$ can be chosen independently of the multiplier sequence $\{\beta_k\}$, and other multiplier sequences like $\{b_k^{(\lambda)}=\exp(2\pi ik\lambda)\}$. For example, in the case of $\{\beta_k\exp(2\pi ik\lambda)\}$, the same set of multipliers $\{b_k=n|B_k|^2/[1+(n-1)|B_k|^2]\}$ will minimize the MISE, and estimated multipliers $\{b_k=[n/(n-1)]-[(n-1)|\hat{B}_k|^2]\}$ will minimize the expected MISE, for any choice of $\{\beta_k\exp(2\pi ik\lambda)\}$. These relationships facilitate computational modularity in the sense that they permit switching between sets of multipliers without altering the multiplier estimation procedure that inputs data into the curve estimation process. Put another way, for trigonometric likelihood functions, and all other curves obtained by substituting one set of multipliers for another set, decisions concerning data usage can be made independently of decisions concerning the particular curve estimated.

8.3 Multiplier system independence

Consider the estimator

$$\hat{g}(\theta) = \sum_{k=-m}^{m} \beta_k b_k \hat{B}_k \exp(2\pi ik\theta)$$

of the curve

$$g(\theta) = \sum_{k=-m}^{m} \beta_k B_k \exp(2\pi i k\theta),$$

where in this section $\{\beta_k\}$ represents any sequence of multipliers chosen prior to data examination. The reciprocal of the sample size n squared times the MISE $J(\hat{g},g)$ equals

$$\sum_{k=-m}^{m} \left\{ n^{-1}(\beta_k b_k)^2 \left(1-|B_k|^2\right) + (1-b_k)^2 |\beta_k B_k|^2 \right\}. \tag{8.3.1}$$

Since and $\mathrm{E}\hat{B}_k = B_k$ and $\mathrm{Var}\hat{B}_k = (1-|B_k|^2)/n$,

$$\sum_{k=-m}^{m} \left\{ (n-1)^{-1} \left[b_k^2 \left(1-\left|\hat{B}_k\right|^2\right) + (1-b_k)^2 \left(n\left|\hat{B}_k\right|^2 - 1\right)\right]\right\} \beta_k^2 \tag{8.3.2}$$

is an unbiased estimator of (8.3.1). For any k such that β_k is nonzero, differentiating either (8.3.1) or (8.3.2) with respect to b_k and setting this derivative equal to zero yields a solution for b_k that is not dependent on any β_k. Thus, $n|B_k|^2/[1+(n-1)|B_k|^2]$ minimizes (8.3.1) and $[n/(n-1)]-[(n-1)|\hat{B}_k|^2]$ minimizes (8.3.2) for any multiplier β_k. Therefore, the choice of the sequence $\{\beta_k\}$ can be made in a modular fashion in that a particular multiplier estimation procedure need not be modified with respect to choice of a curve specifying multiplier sequence like $\{\beta_k\}$.

For example, in trigonometric likelihood applications based on the density $g(z) = Ce^{-c_1|x|-c_2x^2}$, despite the infinite support of the normal and Laplace densities, which over [0,1] are similar in the geometric sense to special cases of f, but are not actually identical to f; the use of the multiplier sequence corresponding to f with c_2 set equal to zero results in a ML estimator $\hat{\theta}$ equal to the sample mean, and with c_1 set equal to zero results in the sample median, as ML estimator $\hat{\theta}$, Tarter (1979a). As described in the previous section, for this reason it may be of value to choose several pairs of constants (c_1,c_2) to construct a variety of different semiparametric regression curves. When this is done there will be no need to re-calculate the sequence $\{b_k\}$ for particular choices of (c_1,c_2).

The form of likelihood where c_1 is chosen equal to zero is also a special case of a general relationship that permits the location parameter ML estimator for any Pearson family model to be determined by finding the point of intersection of two curves, each of which is a special case of $g(\theta)$. However, before proceeding to this general result, we will consider a useful example of multiplier and estimated multiplier coefficients application.

8.4 Sheppard's correction and the lambda method

Suppose $\{b_k\}$ are the Fourier coefficients of the kernel of a generalized histogram estimator $\hat{f}$ of a density f which has well defined moments. Then, relationships given in Section 5.3 imply that an unbiased estimator of the difference between the kth moment of f about the origin, and the integral over the unit interval of x^k times $\hat{f}(x)$ is

$$\sum_{j\neq 0}\left(1-b_j\right)\hat{B}_j\left[(2\pi ij)^{-1}+\sum_{s=1}^{k-1}(-1)^s k!(2\pi ij)^{-(s+1)}/(k-s)!\right]. \qquad (8.4.1)$$

Expression (8.4.1) resembles Sheppard's correction, described on page 69 of Kendall (1943). Sheppard's correction concerns the raw moments $\{\mu_k^*\}$ computed from binned or grouped data, and the true moments $\{\mu_k'\}$ that would have been computed if data had not been binned or grouped. Using the Euler-Maclaurin sum formula the difference between the raw and true moments can be expressed in terms of Bernoulli numbers. The particularly useful special case $\mu_2'=\mu_2^*-H^2/12$ corrects the grouped data second moment about the origin based on the bin width H, the distance between equally spaced consecutive bin centers.

The correction formula $\mu_2'=\mu_2^*-H^2/12$ resembles variance inflation bias relationship $\bar{\sigma}^2=\mathrm{Var}_{\hat{f}}(x)-\mathrm{Var}_f(x)$, in the sense that both can be used to control curve estimator bloating. However, before turning to procedures, which like those based on $\mu_2'=\mu_2^*-H^2/12$, are designed to deal with bias induced by binning specifically, it is informative to consider the following general procedure. To reduce $\overline{\sigma}^2$, the lambda method can be used in conjunction with the special case of (8.4.1) where $k=2$. By means of the lambda method estimators can be constructed that have variances adjusted in order to compensate for bias $\overline{\sigma}^2$. This technique is based on the substitution of $\{b_j \exp(2\pi i j \lambda)\}$ for the sequence $\{b_j\}$ in (8.4.1). When this expression is set equal to zero, a value of λ can be determined that will correct the original estimator for the variance inflation of any nonnegative kernel. This method avoids the construction of estimators that, because of the bloating effects of large $\overline{\sigma}^2$, hide the presence of mixture structure. It can be used iteratively. As a first step, multipliers determined by (8.3.2), or alternatively, the set of weighted MISE-estimated multipliers described in Chapter 4, can be substituted into (8.4.1). Then, the value of λ can be determined that adjusts for variance inflation bias $\overline{\sigma}^2$.

Lambda $\overline{\sigma}^2$ reduction is similar to roughness penalty procedures of penalized likelihood (PL) that counterbalance likelihood against roughness. Unlike the PL however, MISE methodology treats smoothness (in the sense that a nonnegative kernel will tend to smooth an estimator) as a form of bias or distortion. Depending on one's research goals, reality can be seen warts and all or, alternatively, detail can be blurred in order to obtain an overall global impression. The lambda approach to roughness enhancement has one distinct advantage in the sense that visualization of roughness can provide the information needed to trace the actual cause of the anomalies or irregularities.

After a curve's rough structure is traced to its data anomaly source, a researcher has the following two mutually exclusive choices: (1) The data set can be corrected, after which $\overline{\sigma}^2$-neutral procedures can be applied. (2) If data are not remediable, a curve-estimation technique can be selected that is especially designed to deal with the particular data anomaly. For example, semiparametric estimators based on $Ce^{-c_1|x|-c_2x^2}$ can be applied when subgroup membership cannot be assigned to individual data points. In comparison to conventional least-squares parametric procedures, semiparametric M-estimator forms of regression based on $Ce^{-c_1|x|-c_2x^2}$ will be most influenced by that mixture component that has the largest mixing parameter. If, for example, a mixture contained two components with mixing parameters 0.6 and 0.4, often it would be preferable to use conditional switching regression or mode-based regression, which might reveal two distinct regression curves, rather than use a least absolute deviation M-estimator, which would only target the one curve associated with the larger component.

8.5 Correction for grouping

In the sense that it can correct for variance inflation bias $\overline{\sigma}^2$, the λ-based procedure described in the previous section is similar to Sheppard's correction. However, unlike Sheppard's correction, λ-based methodology does not specifically target distortion produced by the binning or grouping of data. We will now describe a procedure that *can* be applied to correct certain ML and other curve estimators for specific forms of data abridgment, binning, and grouping.

As described by Tarter (1979b), within the unit interval for $j=1,\ldots,c$, let $\hat{f}_j$ represent the observed frequency of binned, grouped or abridged data within the jth ordered class interval, x_j the midpoint and h_j the half-width of the jth interval, $x_{c+1}=1$ and $x_0=0$. If the index j is ordered in accordance with the x_j magnitudes, $h_j=(x_{j+1}-x_{j-1})/4)$. The kth unadjusted sample trigonometric moment $\hat{B}_k$ can be expressed as

$$\hat{B}_k = n^{-1}\sum_{j=1}^{c}\hat{f}_j\exp(-2\pi ikx_j). \tag{8.5.1}$$

Suppose the Fejér or some lambda multiplier sequence $\{b_k\exp(-2\pi ik\lambda)\}$ is used to obtain a smoothed estimator f^* from abridged, binned or grouped data. In most cases f^* will not be a satisfactory corrected estimator because the application of a nonnegative Fejér or other bloating $\overline{\text{or}}$ σ^2-increasing kernel, merely substitutes one form of distortion, increased $\overline{\sigma}^2$, for a second form of distortion, bias due to grouping or binning. As the relationship $\mu_2'=\mu_2^*-H/12$ suggests, the latter form of distortion inflates the variance of the underlying estimator. This variance inflation is induced by any nonnegative kernel which, although it may smooth over roughness inherent to grouping, can also substantially increase second moment bias (Bartlett, 1963). Since bias has already been increased due to the grouping or binning process, the direct use of a bloating multiplier sequence or kernel is particularly ill advised.

Consider an alternate to using the single value x_j to represent all $\hat{f}_j$ data points within the jth interval. Specifically, suppose the contribution of the data points within this interval is tallied by integrating the line segment $n^{-1}[\hat{f}_j+\hat{f}_j'(x-x_j)]$ over each half-open interval $[x_j-h_j,x_j+h_j)$, for some suitably chosen $\hat{f}_j'$, such as the derivative of f^* evaluated at x_j. What this process does is replace information that has been concentrated at the class interval midpoint with information spread linearly, but possibly unevenly, over the interval. By using the globally smoothed pdf estimate to evaluate the pdf's derivative at the midpoint of the jth bin, the smoothed estimator f^* determines how the binned data points should be spread within each bin. When all h_j equal the constant Δ, then the group-corrected kth sample trigonometric moment equals

$$\tilde{B}_k = \left[\frac{\sin 2\pi k\Delta}{\pi k}\right]\hat{B}_k + in^{-1}\left[\frac{\Delta\cos 2\pi k\Delta}{\pi k} - \frac{\sin 2\pi k\Delta}{2\pi^2 k^2}\right]\sum_{j=1}^{c}\hat{f}'_j \exp(-2\pi i k x_j). \tag{8.5.2}$$

Note, the coefficient $(\sin 2\pi k\Delta)/\pi k$ is the multiplier that corresponds to the uniform or rectangular kernel, and that this multiplier will itself be the grouping correction coefficient if it happens that

$$\sum_{j=1}^{c}\hat{f}'_j \exp(-2\pi i k x_j) = 0.$$

The following example demonstrates how the binning correction procedure was used to study 42 cases of lung cancer diagnosed at the University of California San Francisco Medical Center in 1970. The solid curve shown in Figure 8.6 is a pdf curve estimator of survival

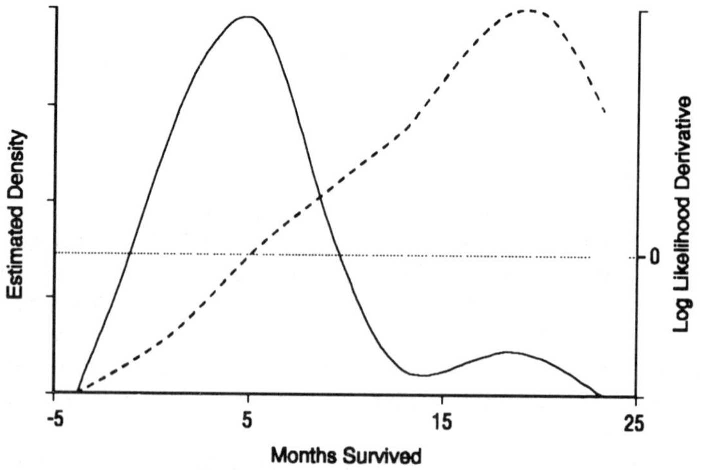

Figure 8.6. *Solid curve is an estimate of the density of the survival times of 42 lung cancer patients. Dashed curve is the derivative of the log likelihood of the location parameter of the normal model.*

time constructed using the procedures described in Chapter 4. The broken curve is the estimate of the log likelihood function's derivative (LLD) obtained using the multiplier sequence based on the standard normal density.

The LLD crosses the x-axis at 5.7 months survival; the sample mean is 5.7381 and the sample median equals 5.0. A curve-based analog to the sample mean, $\int x\hat{f}(x)dx$, equals 5.7. That the sample median equals 5 exactly is an indication of a data analysis anomaly that is present in this, as well as in many other survival analysis applications. Survival information is gathered to the nearest month-survived. For a disease with a relatively optimistic prognosis, for example, breast or colon cancer with localized spread, the average survival time may be sufficiently long to counteract the distortion due to monthly binning. However, for lung cancer studies, where the average survival time may be as short as from 5 to 6 months, abridgment due to binning may be an anomaly of great importance.

Figure 8.7 illustrates the results obtained by executing of the first step of the binning correction procedure. Fejér multipliers were applied to obtain a smoothed estimator. Then the following method was used to determine the constant m of the Fejér multiplier $1-|k|/(m+1)$. The

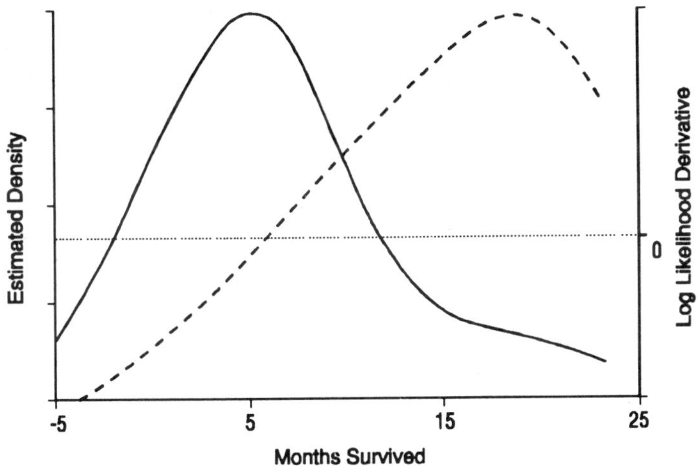

Figure 8.7. *The first step in the binning correction process using the Fejér multiplier system was applied to the curves shown in Figure 8.6 to produce the smoothed pdf and LLD curves shown above.*

subregions under which $\hat{f}$ assumed negative values were identified. These negative estimates, of course, both distort the overall estimator and also reduce the value of $\int(x-\bar{x})^2\hat{f}(x)dx$. Through (8.5.2) the exact value of m was estimated which, when substituted into the kth Fejér multiplier $1-|k|/(m+1)$, increases the variance bias $\overline{\sigma}^2$ of the estimator. The extent to which $\overline{\sigma}^2$ could acceptably be increased was set equal to the increase in estimated variance that would result if $\hat{f}$ were set equal to zero wherever the estimated density assumed negative values.

In general, negative density estimates result from trade-offs between two forms of curve distortion: estimator negativity and variance inflation $\overline{\sigma}^2$. For example, an estimator that exhibits inappreciable estimator negativity invites a very large value of m because, as $m \to \infty$ an estimator based on $1-|k|/(m+1)$ will minimally inflate $\overline{\sigma}^2$. Conversely, an initial estimator that exhibits appreciable negativity suggests that it is advisable to choose a small value of m, one which will substantially smooth a curve estimate, but also increase the bias, $\overline{\sigma}^2$.

Despite the substantial affect that the application of the Fejér multiplier procedure had on the smoothness of the pdf, the LLD curves shown in Figure 8.7 induced very little change in both the ML estimator and in the statistic $\int x\hat{f}(x)dx$. Note that since the calculations for the sample mean and sample median are implemented within the data and not the frequency domain, these statistics will not be altered by the correction procedure. However, the corrected LLD curve shown in Figure 8.7 passes squarely through the sample median rather than the sample mean. (Specifically, after the binning correction process is used, the LLD curve crosses the x-axis at 5.1.) This illustrates how an estimator can be substantially modified by binning correction. Such correction may be very useful in studies of diseases where survival may be of short duration and data is highly abridged (for example, collected in the form of months, rather than days, survived).

8.6 The Pearson and Poisson families

In contrast to the often studied case where the parameters of the Pearson family (PF) general model are estimated using the method-of-moments (Elderton, 1953 page 41), there is very little known about the general problem, as opposed to special cases, of PF parameter estimation using ML methods. Instead, because it is more suitable for conventional ML

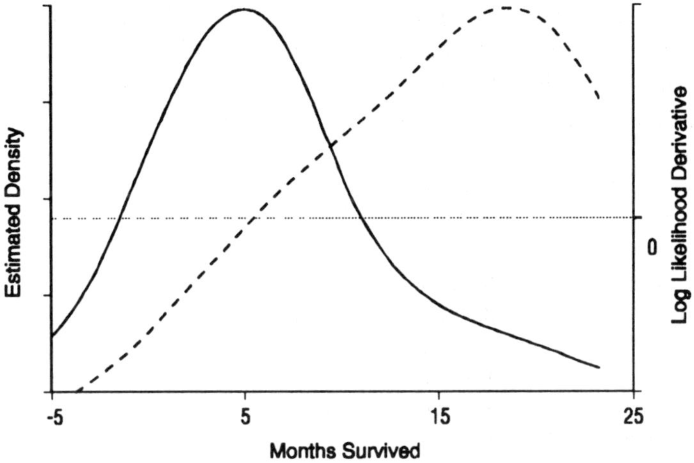

Figure 8.8. *Pdf and LLD curves calculated from lung cancer data after application of complete binning correction procedure. Note that although the pdf has changed substantially compared to Figure 8.6, there is only a slight difference in the LLD curve.*

computation than the PF, the exponential family (EF) generalized model has been given precedence over the PF in much recent statistical literature.

Unlike the conventional likelihood approach, the trigonometric likelihood approach yields compact expressions for all PF location parameter estimates. Specifically, consider the representation of $f(x)=g(x-\theta)$ with location parameter θ, where

$$g'(z)/g(z) = \frac{z\mathrm{I}_{[0,1]}(z)}{(\eta z-\alpha_1)(\eta z-\alpha_2)}. \qquad (8.6.1)$$

As described by Tarter (1983), when $\alpha_1 \neq \alpha_2$, integration by parts yields the ML equation for the likelihood estimator $\hat{\theta}$ of θ

$$\alpha_1 \sum_{k=-\infty}^{\infty} \beta_{1k}\hat{B}_k \exp(2\pi i k\hat{\theta}) = \alpha_2 \sum_{k=-\infty}^{\infty} \beta_{2k}\hat{B}_k \exp(2\pi i k\hat{\theta}), \qquad (8.6.2)$$

where for any index k, β_{1k} and β_{2k} can be evaluated in terms of the sine, cosine, Ci and Si (cosine and sine integral) functions as follows: for v=1,2

$$\begin{aligned}\beta_{vk} &= \int_0^1 \frac{e^{2\pi ikx}}{\eta x-\alpha_v}dx \\ &= \{\cos(2\pi k\alpha_v/\eta)+i\sin(2\pi k\alpha_v/\eta)\} \\ &\quad\times \eta^{-1}\left\{\left(\mathrm{Ci}\left[\frac{2\pi k(\eta-\alpha_k)}{\eta}\right]+i\mathrm{Si}\left[\frac{2\pi k(\eta-\alpha_k)}{\eta}\right]\right)\right. \\ &\quad\left.-\left(\mathrm{Ci}\left[\frac{-2\pi k\alpha_v}{\eta}\right]+i\mathrm{Si}\left[\frac{-2\pi k\alpha_\eta}{\eta}\right]\right)\right\}.\end{aligned}$$

The functions of $\hat{\theta}$,

$\alpha_1\sum_{K=-\infty}^{\infty}\beta_{1k}\hat{B}_k\exp\left(2\pi ik\hat{\theta}\right)$ and $\alpha_2\sum_{K=-\infty}^{\infty}\beta_{2k}\hat{B}_k\exp\left(2\pi ik\hat{\theta}\right)$ are both curve estimators of precisely the form discussed in earlier chapters. Thus, changes in the location parameter estimator attributable to the values of α_1, α_2 and η can be studied by observing the point of intersection of two curves, where each curve is functionally dependent on only one of the two parameters α_1 or α_2.

We have already considered the special normal density case of the Pearson family, which happens to be the $c_1=0$ special case of the family $Ce^{-c_1|x|-c_2x^2}$ upon which LS forms of mean-based semiparametric regression can be formulated. Not surprisingly, this special case yielded multipliers that differed from the multipliers associated with the median-based Laplace $c_2=0$ case of the family $Ce^{-c_1|x|-c_2x^2}$. However, not only will different location parameter estimators yield different series with sample trigonometric moment coefficients, but there can be different series that relate the *same* estimator, for example, the sample mean $\overline{X}$, to the sample trigonometric moments. For example, consider the expression for F^*,

$$F^*(\theta) = \theta-\overline{X}+\frac{1}{2}+(2\pi i)^{-1}\sum_{k\neq 0}\frac{\hat{B}_k}{k}e^{2\pi ik\theta} \quad \text{for } 0\geq\theta\geq 1,$$

which can be unambiguously called the sample cumulative whenever elements of a sample $\{X_j\}$, $j=1,\ldots,n$, satisfy the inequalities,

$1/4<X_j<3/4$ (Tarter and Raman, 1972). If n is odd, setting θ equal to the median M yields the simple relationship for the difference between the mean and median,

$$\overline{X}-M = (2\pi i)^{-1}\sum_{k\neq 0}\frac{\hat{B}_k}{k}e^{2\pi ikM}.$$

Note that were data transformed so that M=1/2, then

$$\overline{X}-M = (2\pi i)^{-1}\sum_{k\neq 0}(-1)^k\frac{\hat{B}_k}{k},$$

so that in this case the more skewed a density f is, the larger will be the sum of the imaginary parts of the $k=1,2,\ldots$ trignonometric moments, each respectively multiplied by $(-1)^k/k$.

Below, for any integer k, the complex number $\hat{B}_k$ will be represented in terms of the two real numbers, $\hat{C}_k$ and $\hat{S}_k$, as $\hat{C}_k+i\hat{S}_k$. Using this notation, by evaluating the Fourier expansion of the sample cumulative F^* discussed in Chapter 2 at $\theta=1/2$, it can be shown that

$$F^*(1/2)+\overline{X} = 1+(2\pi i)^{-1}\sum_{k\neq 0}(-1)^k\frac{\hat{B}_k}{k}$$
$$= 1+\pi^{-1}\sum_{k=1}^{\infty}(-1)^k\frac{\hat{S}_k}{k}.$$

It will be shown in Chapter 10 that in a manner similar to the construction of semiparametric regression curves associated with the normal and Laplace densities, this formula can be applied to construct location parameter ML estimators for logistic model location parameters. An almost identical series can also be obtained by setting θ equal to 0 or 1 in F^*, which yields the simple formulas

$$\overline{X} = \frac{1}{2}+(2\pi i)^{-1}\sum_{k\neq 0}\frac{\hat{B}_k}{k}$$

$$= \frac{1}{2} + \pi^{-1} \sum_{k=1}^{\infty} \frac{\hat{S}_k}{k}.$$

Taking the expectation of both sides of the above formula yields a simple relationship between the population mean and the imaginary parts of the trigonometric moments. That both the population and the sample mean and median have a large number of different relationships with linear expansions in trigonometric moments, suggests many potential new forms of nonparametric regression. Specifically, each individual linear expansion suggests an alternative third step of the three step process in which: (1) the bivariate sample Fourier coefficients are computed, (2) these coefficients are used to estimate the univariate Fourier coefficients of a conditional density, and (3) a point on a regression curve is obtained using the individual formula. Because of this wide variety of alternatives, we will now turn questions concerning the association between trigonometric formulas and the choice of models.

The Poisson model is of particular interest for the following four reasons. (1) As indicated by Hall (1983), useful orthogonal functions can be based on this model. (2) It plays an important role in applied and theoretical statistics. (3) The same statistic, $\overline{X}$, that as shown above has so many relationships with the sample trigonometric moments, is the ML estimator of the Poisson's one parameter. (4) In a crucial way, series associated with truncated forms of the Poisson model are fundamentally different from series associated with trunctated normal and Laplace models. These series yield exact relationships yet involve finite, rather than infinite, sums. Since, in practice, all data must be input with a finite degree of precision, these sums have several interesting applications.

The simplicity of PF location parameter curves depends, in large part, on the assumption that the support of the density is on the unit interval. This is ensured by the $I_{[0,1]}(x)$ term of (8.6.1). A similar form of truncation can be applied to Poisson densities. In addition to truncation, there is another aspect of the Poisson density of a discrete random variate that is relevant to applications of likelihood multipliers.

Although continuous densities are commonly parametrized in terms of some location parameter, expressions involving location parameters are not usually encountered when discrete models are studied. However, as if to compensate for the complexities brought about by the lack of location parameter representation, discrete density Fourier expansions have a very useful feature. While, in the continuous variate case, infinite expansions are needed to represent most models as Fourier series, infinite expansions are not required for representation of curves associated with discrete variates.

Specifically, with one very minor modification the $\{\exp(2\pi ikx)\}$ system can be applied to obtain useful estimates of a discrete random variate's density. As described by Lanczos (1956, page 233), we will use the symbol $\sum_{s=0}^{m}{}'$ to represent a sum whose zeroth and mth terms are half the value of the values assumed by the usual sum operator. This modification allows the combined contribution of the first and last sum terms to equal that of other individual terms. The process of merging first and last sum terms is analogous to connecting the endpoints of the circle on a clock-face, where the integer 12 shares the same space as the value 0. For example, for an integer valued v,

$$\sum_{s=0}^{m}{}' e^{2\pi isv/m} = \frac{1}{2} + e^{2\pi iv/m} + \ldots + e^{2\pi i(m-1)v/m} + \frac{1}{2} \tag{8.6.3}$$

which, for $0<v<m<\infty$, equals $\exp(\pi iv)\sin(\pi v)\cot(\pi v/m)$. Consider the sequence $\{\Psi_k(s/m)\}$ where $k=1,2,\ldots,m$, and

$$\Psi_k(s/m) = (\sqrt{m})^{-1} e^{\frac{2\pi iks}{m}}. \tag{8.6.4}$$

Then, (8.6.3) implies that

$$\sum_{s=0}^{m}{}' \Psi_k(s/m)\Psi_j^*(s/m) = \delta_{k,j}, \quad \delta_{k,j}=0, \text{ if } j\neq k; \ =1, \text{ if } j=k \tag{8.6.5}$$

where Ψ_k^* represents the complex conjugate of Ψ_k.

For the discrete variate density $f(x|\mu)$, suppose

$$B_k = \sum_{s=0}^{m}{}' \Psi_{-k}(s/m) f[(s/m)|\mu] \tag{8.6.6}$$

and let the estimator of the kth Fourier coefficient of $f(x|\mu)$ be defined as

$$\hat{B}_k = \sum_{s=0}^{m}{}' \Psi_{-k}(s/m)\hat{f}(s/m), \tag{8.6.7}$$

where $\hat{f}(s/m)$ is the observed frequency of the observed value s/m. Then basic relationships for discrete variate Fourier expansion described by Lanczos (1956, p. 234) imply that

$$\log f((s/m)|\mu) = \sum_{k=0}^{m}{}' \Psi_k^*(s/m) \sum_{v=0}^{m}{}' [\log f((v/m)|\mu)]\Psi_k(v/m). \tag{8.6.8}$$

It follows that the log likelihood (logL) of an iid discrete valued sample $\{X_j\}$ can be expressed as the weighted sum

$$n^{-1}\log L = \sum_{k=0}^{m}{}' \hat{B}_k w_k(\mu), \tag{8.6.9}$$

where

$$w_k(\mu) = \sum_{s=0}^{m}{}' [\log f((s/m)|\mu)\Psi_k(s/m).$$

We will now proceed to apply (8.6.9) to obtain several useful formulas for the ML estimator of the Poisson parameter μ.

As previously discussed, a density $f(x)=g(x-\theta)$, where $g(z) = C\exp[-c_1^2z^2]$, has as its location parameter θ's ML estimator, the sample mean, $\overline{X}$. Yet the ML estimator of a Poisson parameter μ is also the sample mean, $\overline{X}$. In Chapter 10 will consider the question: Whether there are conditions which ensure that there is a one-to-one correspondence between standard models and ML parameter estimators? Yet even if there were to be a one-to-one correspondence between models and location parameter ML estimators, here it is not redundant to consider the estimation of Poisson parameter μ, which is not a location parameter, separately from the estimation of μ of $g(x-\mu)$, because one of the most interesting of Fourier series peculiarities. Specifically, different equations for $\overline{X}$ in terms of the sample trigonometric moments are associated with the Poisson model and the normal model. Unlike the normal-based equation, the Poisson equation is a finite expansion or a trigonometric polynomial. But this equation only is valid for the sample mean of a discrete valued sample. Yet all computer data input is discrete to a degree determined by either single or double precision routines. Consequently, in practice the finite weighted sums discussed below can be made to yield exact relationships for regression and other applications.

To clarify the relationship between data degree of precision and the exactness of likelihood multiplier representation, the following experiment was performed: This experiment was designed to show the manner in which two different linear sums of sample trigonometric moments, one based on $\sum_{k=0}^{m}{}' \hat{B}_k w_k(\mu)$, which equals $\overline{X}$, and a second based on the standard model $g(z) = C\exp[-c_1^2 z^2]$ and the relationship $\beta_k(\theta) = \beta_k \exp(2\pi i k\theta)$, differ. The abscissas of each graph shown in Figures 8.9 through 8.12 record the number of terms used in a truncated representa-

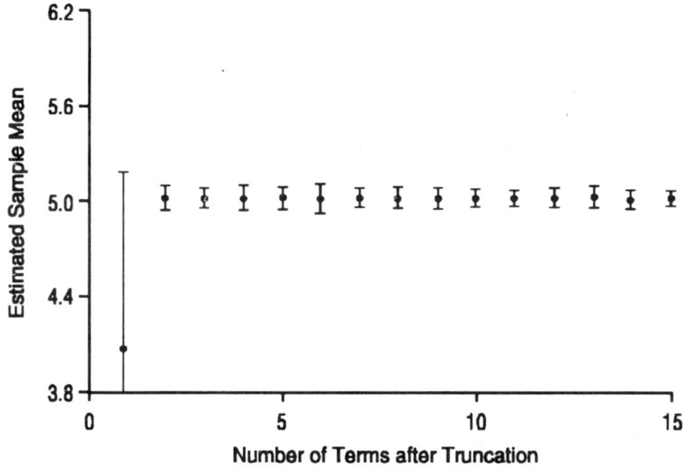

Figure 8.9. *Truncated series approximation to the sample mean based on the normal model. The data were Normal with* $\mu=5$ *and* $\sigma=1$. *Each point is the average of 50 trials where* $\overline{X}$ *was estimated by truncated series approximation from 10 sample points. The error bars encompass* ± 2 *standard deviations around each average.*

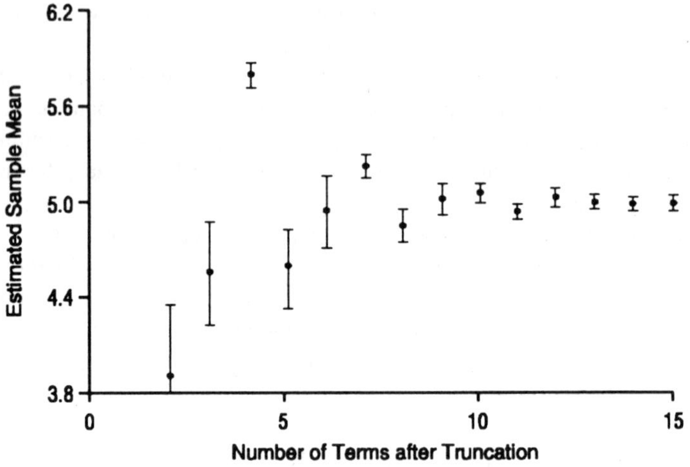

Figure 8.10. *Truncated series approximation to the sample mean based on the continuous Poisson model. The data were Normal with* $\mu=5$ *and* $\sigma=1$*. Each point is the average of 50 trials where* $\overline{X}$ *was estimated by truncated series approximation from 10 sample points. The error bars encompass* ± 2 *standard deviations around each average.*

tion of $\overline{X}$. The trigonometric likelihood expansions whose performance is described by Figures 8.9 and 8.11 were based on the standard normal model $g(z) = C\exp[-c_1^{\ 2}z^2]$. Those whose performance is described by Figures 8.10 and 8.12 were based on the Poisson model.

Two types of data were used to compute the sample trigonometric moments. Figures 8.9 and 8.10 describe the computation of $\overline{X}$ based on normal data where the population mean was chosen to equal 5 and the population standard deviation was chosen to equal 1. Figures 8.11 and 8.12 describe the computation of $\overline{X}$ based on conventional Poisson data where the parameter μ of the Poisson model was defined to equal 0.2. This Poisson parameter was chosen to match Poisson and normal estimated densities as closely as possible, when these curves were transformed to the unit interval as part of the computational routines used to conduct the comparison studies.

Sample sizes $n=100$ were used throughout the experiment. The boxplots of the four figures were based on 50 independent trials per figure, for a total of 200 trials based on a total of 10,000 normal and 10,000 Poisson variates. The box edges were set equal to the .025 and .975 sample quantiles.

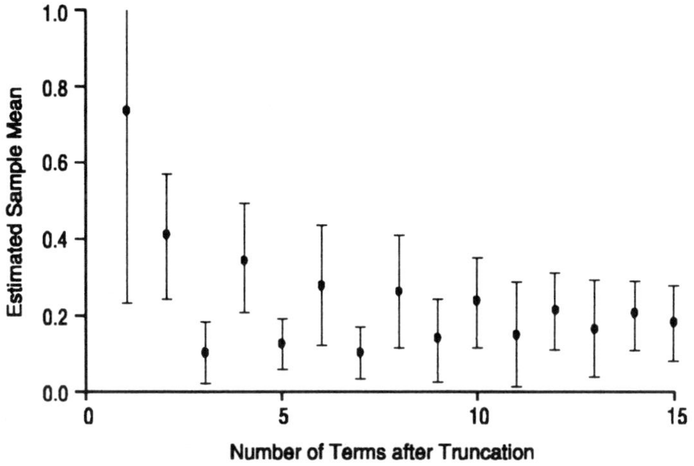

Figure 8.11. *Truncated series approximation to the sample mean based on the normal model. The data were Poisson with* $\lambda=0.2$*. Each point is the average of 50 trials where* $\overline{X}$ *was estimated by truncated series approximation from 10 sample points. the error bars encompass* ± 2 *standard deviations around each average.*

Figure 8.9 clearly shows that even when only two trigonometric moments are used to approximate the sample mean of normal variates, if the appropriate likelihood-based expression is used, the series converges fast enough to provide useful estimators. Since each series converges to the sample mean, and not the population mean of 5, it is not surprising that the bands do not converge to a point. The variance of the *k*th sample trigonometric moment equals $(1-|B_k|^2)/100$. Because $|B_k|^2$ decreases rapidly as k increases, the corresponding increase in $(1-|B_k|^2)$ accounts for the slight variations in the box-plot widths when different numbers of terms are retained after truncation.

After twelve terms, the $\overline{X}$ series described by Figure 8.10 converges to yield a box-plot that is almost identical to the box-plot shown in Figure 8.9. However, the actual series representation of $\overline{X}$ was derived by assuming the Poisson model described below. Consequently, the wild fluctuations of the box-plots shown above the number-of-terms axis are due to a mismatch between the density model and the likelihood model.

(We have found that slow fluctuation damping only is important

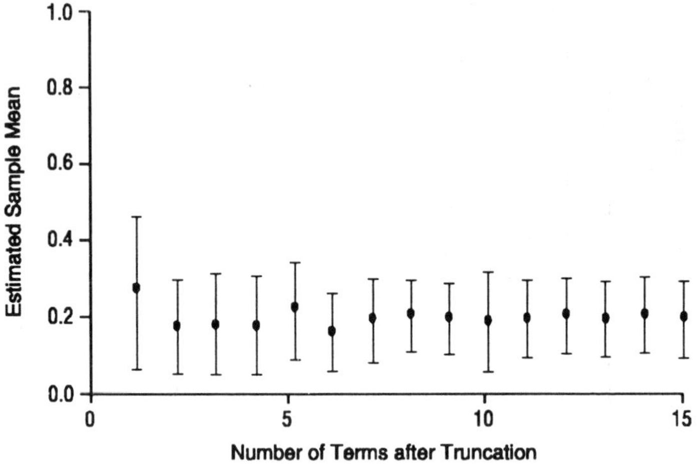

Figure 8.12. *Truncated series approximation to the sample mean based on the continuous Poisson model. The data were Poisson with* $\lambda=0.2$*. Each point is the average of 50 trials where* $\overline{X}$ *was estimated by truncated series approximation from 10 sample points. The error bars encompass* ± 2 *standard deviations around each average.*

when lambda or likelihood multipliers are used with small size samples. For these samples the requirement that many series terms be used to achieve convergence may substantially limit the applicability of contrast modification and semiparametric regression methodology.)

Use of the same series expression that performed so well for normally distributed data to compute $\overline{X}$ for a Poisson sample resulted in the trials described by Figure 8.12. Here, the oddness or evenness of the number of series terms was related to convergence which, on the whole, was not as rapid as for normal data. However, as shown in Figure 8.12, when the Poisson $\overline{X}$ formula was used with Poisson data, similarly to the series described by Figure 8.9, the series performed almost as well for two trigonometric moments as it did for ten terms. Although there was some change in box-plot width as the number of terms decreased, the magnitude of this change was less dramatic than the magnitude of the difference between the one term and the two term expansion.

The results of the above study have the following practical implication: For moderate to large size samples, when there is considerable doubt as to model validity, as there may be in regression applications where very little is known about the form of the distribution of residuals about the regression curve, it is advisible to use many series terms to estimate a location parameter rather than use a stopping rule that leads to heavy truncation. Also, if lambda or other multipliers are used in conjunction with trigonometric representation, particular care should be exercised in the choice of series used to estimate a location parameter. By choosing the appropriate series, only a small number of terms may be required and, as a consequence, the resulting series will yield accurate estimators from very short sequences of sample trigonometric moments.

As described in Section 8.2, it is possible to devise nonparametric regression procedures so that data influences the regression curve or curves through estimates of the Fourier coefficients $B_\nu^{(x_0)}$ of

$$\sum_{\nu=-\infty}^{\infty} B_\nu^{(x_0)} e^{2\pi i \nu y}, \quad \text{where } B_\nu^{(x_0)} = \left[\sum_{\nu=-\infty}^{\infty} B_{s,\nu} e^{2\pi i s x_0} \right] \Big/ f(x_0)$$

for a sequence of estimated conditional densities $f(y|x)$. An alternative formula for the sample mean, which provides an alternative way of finding a measure of central tendency based on the estimated Fourier coefficients of the above sequence of conditionals, can be of considerable value.

For example, in much the same way that least LAV and LS semiparametic regression curves can be based on the Laplace and normal density models, respectively, a form of regression based on the Poisson model can be devised. Since $\overline{X}$ is the common ML estimator for both the normal and Poisson case, regression bands based on the Poisson model will approach the same estimates as will LS semiparametric regression, as the number of series terms increases. However, when LS is actually implemented, truncated series representation must be used. Such a series does not simply provide an approximation of the desired regression curve. Instead, the desired regression curve is obtained both by truncation, and through the use of the multiplier procedures that determine the crown and the brim of some KT . Thus, Poisson-based Fourier representation of the sample mean can provide a useful regression alternative when data of only limited precision is available.

In summary: there are two likelihood multiplier-based semiparametric regression alternatives. (1) The grouping correction procedures described in Section 8.3 can be used to modify estimators of the conditional density's Fourier coefficients. These corrected estimators can be used in conjunction with LAV, LS and other regression procedures to form a semiparametric curve that has its own advantages and disadvantages. (2) Alternatively, in some cases the Poisson form of $\overline{X}$ representation can be used directly to deal with discrete-ized data.

In order to compare results over the same [0,1] support region used for the Pearson-like model, it is convenient to introduce the truncated and transformed density

$$f_p(z|\mu) = me^{-\mu}C_1(\mu)\{\mu^{mz}/(mz)!\}I_{[0,1]}(z), \tag{8.6.10}$$

where $C_1(\mu)$ and μ are constants chosen to ensure that for the size n iid sample $\{Z_j\}$ $j=1,\ldots,n$: (1) if a variate Z_j with density $f(z|\mu)$ is multiplied by the integer m, then the resulting variate will have a density geometrically similar to a Poisson density with parameter $\tilde{\mu}$; (2) the sum of f evaluated at 0, $1/m$, $2/m$,... equals 1; and (3) mZ_j is an integer for $j=1,\ldots,n$. Defining

$$\{\Psi_k(s/m)\} = (\sqrt{m})^{-1}\exp(2\pi ik(s/m))$$

as the kth member of the orthogonal sequence $\{\Psi_k(s/m)\}$, it follows that the derivative

$$d\log f(z|\mu)/d\mu = -1+C_1'(\mu)/C_1(\mu)+zm/\mu,$$

where $C'(\mu)$ is the derivative of $C_1(\mu)$ with respect to μ. Thus, the derivative of the kth Fourier coefficient

$$d\beta_k(\mu)/d\mu = \sum_{s=0}^{m}{}'[-1+C_1'(\mu)/C_1(\mu)+s/\mu]\Psi(s/m)$$

for $k \neq 0$ and

$$d\beta_0(\mu)/d\mu = [-1+C_1'(\mu)/C_1(\mu)]\sqrt{m}+(m+1)\sqrt{m}/2\mu.$$

The relationship

$$\left(\sqrt{m}\right)^{-1}\sum_{s=0}^{m}{}' s\exp(2\pi i(s/m)k = -(i/2)\sqrt{m}\cot(k\pi/m)$$

yields the following formula for the MLE $\hat{\mu}$ of the μ parameter of $f_p(z|\mu)$:

$$\hat{\mu} = (m+1)/(2\sqrt{m}) - S(m), \tag{8.6.11}$$

where

$$S(m) = \sum_{s=0}^{m}{}' \operatorname{Im}(\hat{B}_k)\cot(k\pi/m)$$

if m is even, and

$$S(m) = \sum_{k=1}^{(m-1)/2}{}' \operatorname{Im}(\hat{B}_k)\cot(k\pi/m)$$

if m is odd, and where if $\hat{B}_k = \hat{C}_k + i\hat{S}_k$, $\operatorname{Im}(\hat{B}_k) = \hat{S}_k$.

In practice, when formula (8.6.11) is applied, m should be set equal to at least one *plus* the largest observed Poisson variate value. If this is not done, the first and the last estimated frequencies, $\hat{f}(0)$ and $\hat{f}(1)$, may not be equal. When $\hat{f}(0)$ and $\hat{f}(1)$ are not equal, (8.6.11) will merely be an approximation and not, as it is otherwise, identically equal to $\overline{X}$. Formula (8.6.11) which expresses $\overline{X}$, the ML estimator of the μ of $f_p(z|\mu)$, is an expansion whose coefficients are the imaginary components of a *finite* number of the sample trigonometric moments. Although this formula is, in theory, only valid for discrete data, in practice it is always applicable, with the important proviso that care is taken to set the number of series terms in the correct correspondence with the degree of data precision.

Despite use of truncated model (8.6.10), rather than a Poisson model with infinite support, likelihood expression (8.6.11) is an extremely accurate formula for $\overline{X}$ calculated from discrete samples yet, because of the above proviso, it is not particularly accurate for continuous data.

For example, 50 trials, each consisting of 100 iid normal N(μ,1) variates enumerated using double precision, were conducted for each choice of mean μ=0.2, 0.5, 1.0, 2.0 and 5.0. In each trial, the actual value of $\overline{X}$ was calculated directly and by (8.6.11) and the difference between these two statistics determined. For the five choices the averages and standard deviations of each 100 differences were (1.22,0.156), (0.763,0.119), (0.423,0.097), (0.076,0.041), (0.029,0.021), respectively. However, when the same choices were used to conduct trials with Poisson data, all averages and standard deviations were all less than 0.0000006. This illustrates that, at least in the univariate case, choices between multiple trigonometric expressions for the same statistic should be made on the basis of the form of data used. In the next section we will discuss a second difference between discrete and continuous variates, which concerns the connection between the number of estimated probabilities and the number of coefficients or parameters.

8.7 The Neyman discrete variate test

One of the first uses of generalized representation for statistical inference involved a variant of the truncated form of the Poisson model discussed in the previous section (Neyman 1946 p. 159). This application was based on chi-square test procedures and not on likelihood methodology. Nevertheless, it is informative to consider how Neyman devised his test as an introduction to the likelihood ratio procedures described in the next chapter.

According to Neyman, if there is no a priori information concerning the probabilities P{Y=y}, of a discrete valued random variable Y, it can be assumed that the set of admissible hypotheses contains every hypothesis ascribing some specified values to the probabilities

$$p_y,\ y=0,1,\ldots,n-2, \text{ and } p_{n-1} = 1-\sum_{y=0}^{n-2} p_y$$

and therefore these probabilities can be used as $n-1$ independent parameter components of the vector θ.

Neyman was foresightful to make this connection between probabilities and what he called parameters or, as discussed in Chapters 6 and

7, curve properties. However, this approach is limited in the sense that it cannot be directly extended to the continuous variate case. Unlike Poisson and other variates whose support regions are denumerable, a continuous variate has a nondenumerable support region. This makes it impossible to determine the numerical difference between the number of parameters specified by the null and the alternative hypotheses. This difference is crucial both to Neyman's chi-square and to likelihood ratio (LR) procedures.

If the support region of a variate Y is denumerable, the cdf of the distribution must be a step function and Y cannot be a continuous random variate. It follows that the support region of a continuous random variate contains a nondenumerable number of distinct values. However, the assumption that f has a uniformly convergent Fourier expansion implies that there are a denumerable number of curve properties, the Fourier coefficients $\{B_k\}$, which determine f. Therefore, as will be discussed further in Chapter 10, the sequence $\{B_k\}$ can be used for inferential purposes in much the same way that Neyman used the set of all probabilities of a discrete denumerable-valued Poisson variate.

Many different procedures can be used to construct inferential procedures based on $\{B_k\}$. One technique, which is a direct extension of Neyman's Poisson procedure, will be described in the next chapter. Since the estimates of $\overline{X}$ provided by (8.6.11) are ML estimates, they can be used to construct approximate likelihood ratio tests for a Poisson response variate Y conditioned on some selected value of the key variate x_j. For example, suppose that the density $f(y|x)$ of the variate, months survived, given a treatment regimen coded by x_j, can be described very well by the Poisson model. Using methods described in the next chapter, the Fourier coefficients of $f(y|x)$, for a given value of $x=x_j$, can be estimated and used in conjunction with (8.6.11) to determine the conditional analog of univariate $\overline{X}$, $\overline{Y}|x_j$. The Neyman procedure or its ML counterpart can then be used to test hypotheses concerning the true Poisson parameter μ.

8.8 Discrete variates and logistic regression

This section describes procedures that use a nonparametric estimator of $f(y|x_1,x_2,\ldots,x_q)$, where y is the value assumed by a discrete variate Y, to check, and in some instances, circumvent, the assumptions that underlie logistic regression. Logistic regression is commonly used to estimate the relationship between the log odds ratio, the log of the chance of having a disease over the chance of not having a disease, and a linear combination of independent variates.

The event that a subject has, or does not have, a disease, is a binary valued variate that could be conceptualized as a two-valued special case of the discrete valued variables discussed in the previous two sections. However, to ensure that the estimated density takes on equal boundary values, it is convenient to apply a four-valued discrete formulation of the binary case. In terms of the equally spaced values, 0, 1/3, 2/3, 1, $f(s/m)=0$ when $s=0$ and m, and the numbers 1/3 and 2/3 are used to represent the values of the binary variable within (8.6.7), which are usually coded as 0 or 1 (Ott and Kronmal, 1976; Tarter, 1986).

There is a useful way in which a curve estimator of the density $f(y|x_1,x_2,\ldots,x_q)$ can be applied. An estimate of the odds ratio, $f(1/3)/f(2/3)$, can be used to determine a plotting character in a manner similar to the representation of estimated probabilities by specific plotting characters that was described in Section 5.6. For example, suppose points corresponding to patients whose measurements are associated with high odds of having a disease are displayed as small squares, while individuals whose measurements are associated with low odds of having this same disease are displayed as using small circles. Specifically, estimated odds ratios greater than 2 correspond to squares and ratios less than 1/2 correspond to circles. In addition, any point that corresponds to an odds ratio between 1/2 and 2 is not shown, which will result in a blank band between the (>2, small squares) and (<1/2, small circles) bands.

The borders between either the blank and the square-symbol or the blank and circle-symbol band, provide considerable information concerning the logistic regression relationship. The conventional representation of the logistic model

$$\begin{aligned}\text{logit}(p) &= \log[p/(1-p)] \\ &= \alpha+\beta_1 x+\beta_2 x+\ldots+\beta_q x\end{aligned} \tag{8.8.1}$$

can be more clearly expressed as

$$\log\left[\frac{f(y|x_1,x_2,\ldots,x_q)}{1-f(y|x_1,x_2,\ldots,x_q)}\right] = \alpha+\beta_1 x+\beta_2 x+\ldots+\beta_q x. \tag{8.8.2}$$

For example, suppose $y{=}2/3$ indicates the presence of a symptom, and that the variables $X_3,..,X_q$ take on a single set of values, $x_3^{(c)},\ldots,x_q^{(c)}$. If

$$K = \log\left[\frac{f(y|x_1,x_2,x_3^{(c)},\ldots,x_q^{(c)})}{1-f(y|x_1,x_2,x_3^{(c)},\ldots,x_q^{(c)})}\right] \tag{8.8.3}$$

for some pair x_1,x_2, where C is the constant value, $C = \beta_3x_3^{(c)}+\beta_4x_4^{(c)}+\ldots+\beta_px_p^{(c)}$, it then follows that

$$K-C = \log\left[\frac{f(y|x_1,x_2,x_3^{(c)},\ldots,x_q^{(c)})}{1-f(y|x_1x_2,x_3^{(c)},\ldots,x_q^{(c)})}\right]-C \tag{8.8.4}$$

is constant for the point pair x_1 and x_2. Therefore, the assumption of a logistic model implies that $\alpha+\beta_1x+\beta_2x = K-C$, which in turn implies that for the single given sequence $x_3^{(c)},\ldots,x_q^{(c)}$, (8.8.4) is an implicit formula for a linear relationship between x_2 and x_1.

If the regression equation contains only two variates, then this geometric interpretation of the logistic model can be used directly. For higher dimensions, two nonparametric density estimators, the first from the subpopulation for which $y{=}1/3$ and the second for which $y{=}2/3$, can be obtained. Then some suitable single value of the third through qth variates selected and, finally, the equi-value contours of (8.8.3) can be used to check on the validity of the logistic model.

Suppose, for example, that the two variate regression function, $\alpha+\beta_1x+\beta_2x$, applies. If $\beta_1{=}0$, contours will not only be linear but will parallel the x_1 axis. On the other hand, if $\alpha+\beta_1x+\beta_2x$ is not a valid form of representation, contour nonlinearity indicates that variate values should be transformed, for example, to yield $\log x_1$ and $\log x_2$, and the new set of independent variate values checked in the same way that the old values are checked. In this way the particular combination of key variates that best predicts the disease symptom response variate can be determined.

CHAPTER 9

Survival curve and bivariate estimation applications

While the sources of information used for conventional data-based likelihood differ from those used to implement the trigonometric (frequency domain) likelihood techniques described in the last chapter, both classes of procedure are designed to estimate a finite number of parameters. Although the methods differ in the sense that information is input differently, their goal is the same. We now will introduce a form of generalized likelihood which has as its end the solution of a problem that differs considerably from the estimation of a model's or an extended model's parameters.

Consider that both the mixture decomposition process discussed in Chapter 5 and the v-model parsing process discussed in Chapters 6 and 7 concern a form of semiparametric representation which involves two or more components. For example, in the case of the mixture $f(x|p,\mu_1,\mu_2,\sigma) = \{pf_1([x-\mu_1]/\sigma) + (1-p)f_2([x-\mu_2]/\sigma)\}/\sigma$ these components are f_1 and f_2, while in the case of the v-model $G\{(\mathrm{v}[x-\mu_1]-\mu_2)/\sigma\}$ these components are v and G. Much of this chapter also concerns a problem that involves two distinct components. However here, rather than the need for a decomposition or parsing process to disentangle components, data point attribution to some specific component is known a priori. One way to interpret this distinction is to compare the problem discussed later in this chapter to one particular application of mixture methodology, as follows.

Step 4 of the mixture decomposition process described in Section 5.6 consists of the separation of information in the frequency domain into individualized subsequences, where each subsequence is associated with a particular mixture component. After this separation is accomplished individual multiplier sequences $\{b_k^{(1)}\}$ and $\{b_k^{(2)}\}$ can be calculated; where each sequence corresponds to an individual component and is optimized as described in Chapters 3 and 4 by using either the unweighted, $w(x)\equiv 1$, or weighted MISE metric.

If, rather than being interested in the individualized mixture components themselves, a researcher wishes to obtain an improved estimator of the entire pdf of the mixture, the individualized components can be reassembled. The mixture reassembly or reconstitution process resembles the procedures used to construct population life tables. In most life tables a short, commonly year-by-year, degree of abridgment is applied for early years and a longer, usually five year, form of class interval grouping is applied to input data concerning subsequent years. This use of a variable abridgment or variable class interval size is in accord with the greater change in mortality rates immediately after birth and the more gradual change in mortality that occurs for five year old and older individuals. For the mixture-based variable kernel method, two or more different component specific multiplier sequences can be used, where each sequence corresponds to a different, for example, yearly and five-year, degree of abridgment. Typically, the kernel that corresponds to a mixture component that has a small second cumulant will have a higher peak than will the kernel that corresponds to a component that has a moderate to large second cumulant.

The variable abridgment of a population life table and the variable multiplier systems, or variable KT's or kernels (Breiman, Meisel and Purcell, 1977), of mixture reconstitution differ in the sense that, while for the life table information is modularized in the data domain, for the mixture, information is modularized in the frequency domain. The allocation of individual data points to individual modules is circumvented by separating the sample trigonometric moments calculated from the entire sample into two or more modules, $\{\hat{B}_k^{(1)}\}$ and $\{\hat{B}_k^{(2)}\}$, from which the multipliers $\{b_k^{(1)}\}$ and $\{b_k^{(2)}\}$ are individually determined.

While the life table and the mixture procedure differ in their respective use of data versus frequency calculations they do share at least one important characteristic. Once modules are determined, there is little or no question concerning the reassembly process by which these modules are used to construct a curve estimator. On the other hand, it is the reassembly process where the methods described below make use of likelihood methodology. The formula used to form a composite from individual modules is suggested by examination of the trigonometric likelihood function.

Section 9.2 describes a method for forming a composite survival curve based on subsamples of both complete and incomplete (censored) observations. The relationship upon which this method is based was derived by substituting an m-term Fourier expansion into the Bartholomew (1957) likelihood function, BLF. The BLF is the joint probability of obtaining the observed known-complete survival times as well as the so-far-survived measurements of individuals who are still alive at the date of study completion or other endpoint of the period of observation. This later subpopulation is usually characterized by the term "incomplete," a term that will be discussed in detail in Section 9.3. Equations based on the BLF were used by Feigl and Zellen (1965) as well as Zippin and Armitage (1966) in the exponential model case.

The BLF was used by Tarter (1978, 1979a) in a manner similar to applications to the exponential model case with the exception that an m-coefficient Fourier expansion replaced the single parameter exponential model. In this way a formula was obtained for the reassembly of modules consisting of what were characterized by Peterson (1977) as "subsurvival functions." Specifically, the subsurvival function of the completes equals the module: one minus the cdf, i.e., $1-F_C$, of the population from which complete survival measurements are drawn; while that of the incompletes is $1-F_I$ where F_I corresponds to the population of incompletes. Hence, here the question arises: How can the modules $1-F_C$ and $1-F_I$ be used to construct the composite survival curve $S(t)\equiv 1-F(t)$, where F is the cdf of the distribution of survival times of all individuals who are under study, the duration of the study notwithstanding.

Despite the similarity of the approach based on Fourier series and approaches based on the exponential elementary function model, as far as mathematical derivation is concerned, the applications of the generalized representation and elementary function approaches are very different. Unlike survival analysis based on an exponential model, or some other elementary function, the application of the generalized methods to the survival curve or hazard function estimation problem leaves considerable room for the selection of individual modules. Like the mixture decomposition-reconstitution form of variable kernel pdf estimation, individual sets of multipliers, corresponding to two or more possibly distinct kernels, can be applied.

The advantage of using likelihood to determine the composite reconstitution process, and then using separate procedures to construct modules from the individualized complete and incomplete subsamples, stems from the distinction between the use of metrics like MISE for curve estimation purposes and the use of a two-valued loss function in conventional frequentist forms of inference. Consider, for example, a typical clinical trial application. For such an application it is much more important to assess accurately long-term survival outcomes of the drug or treatment under study than it is to estimate a survival curve or a hazard function evaluated close to the point $t=0$, where the value 0 designates date of diagnosis or date of initial treatment. Hence, the MISE weighting procedures described in Chapter 4, the equivalent weighted absolute value metric, or the corresponding penalized likelihood alternatives, can all provide the tools necessary to input a crucial form of information. Specifically, that it is in the right tail that accuracy is desired, at the possible expense of local estimator reduction in efficiency near $t=0$.

The subsurvival-based use of likelihood acts as a springboard or starting point for more complex methodologies. This is similar to penalized likelihood (PL) methods which involve the subtle interplay of conventional likelihood with smoothness considerations or roughness penalties. For the generalized procedures described below which use the Bartholomew (1957) likelihood equations, smoothness is controlled by the MISE metric. But, this metric can be modified in a manner that does not necessarily have much to do directly with smoothness or roughness. For example, it can be used to enhance efficiency in the right tail of estimators (to better judge long term survival) at the acceptable or remediable expense of efficiency loss in the midportions and left side of the survival variate's support region.

The above approach does not correspond to any application of conventional parametric likelihood, either when used in conjunction with the two-valued loss framework of frequentist hypothesis testing, or the all, or nothing, inclusion inherent in the use of confidence bands about an estimated curve. In the latter case, it is quite likely that bands would narrow at some point t_0 in accordance to an increase in the user selected value of MISE weight $w(t)$ which assumes a large value at point t_0, but the degree and exact interpretation of this relationship has yet to be determined.

Before turning to applications of trigonometric likelihood that are primarily based on MISE, for purposes of completeness it is appropriate to further discuss the penalized likelihood, PL, approach to curve estimation. PL methodology supplements the conventional log-likelihood function L with a roughness penalty β, which is multiplied by the measure of density roughness R. The composite function $\mathrm{L}-\beta R$ is then maximized, much in the same way that the traditional maximum likelihood, ML, function is maximized in the parametric case.

That the roughness penalty approach balances representational complexity and data availability is elaborated upon in a commentary on the work of Good and Gaskins (1980) by one of the authors paraphrasing Keats' as follows: "Beauty is truth, truth beauty, that is all/ye know on earth, and all ye need to know,' say Good and Gaskins: *not so,*" (Tarter, 1980). β serves as a damping factor that counterbalances truth, in the form of likelihood, against beauty, in the sense that the smooth and regular is beautiful. Before the advent of generalized statistical approaches, smoothness was ensured by limiting the choice of representation to elementary functions, like the normal or lognormal. Unlike the model-based approach, both PL and MISE methods can deal with the problem of roughness directly, and need not merely assume it away.

The MISE-based methodology concentrated upon in this book deals with the possible over-fitting of a generalized representation by means of the approaches discussed in Chapters 3 and 4. It relies on the premise that inefficiency (specifically the excess expected integrated squared error of overfitted representation) can provide the same damping which can also be achieved by modifying the PL's user-supplied constant β. As described in Chapter 5, smoothness *can* be enhanced by using a negative choice of the lambda method λ or a small integer value of the Fejér method's constant m. But, when such choices are made it is recognized that this artificially induced smoothness is a form of bias, useful in some applications, but nonetheless, still a distortion.

9.1 Likelihood in the generalized case

In this section we will show that the sample trigonometric moments $\{\hat{B}_k\}$ are unbiased ML estimators of the population trigonometric moments $\{B_k\}$. Many properties of a density, such as the mean, can be expressed as linear functions of members of the sequence $\{B_k\}$. Hence, since $\{\hat{B}_k\}$ are ML estimators, the maximum likelihood estimators of many statistics can be calculated by means of these linear functions.

A considerable portion of this section will make use of the theorem that any continuous density f that has a uniformly convergent Fourier series expansion is uniquely determined by that expansion (Edwards, 1967, page 41). This and other variants of Fourier series uniqueness theorem, such as that outlined in the next paragraph, are not only basic to the proof that $\{\hat{B}_k\}$ are ML estimators but, as outlined below, can also be used to derive ML estimators of the survival curve S, which are more useful in many applications than are the Kaplan Meier and life table survival curve estimators.

Sz-Nagy (1965, page 82) asserts that any continuous function of period 2π can be approximated uniformly, as accurately as we wish, by trigonometric polynomials. He then applies the Weirstrass-Stone theorem in conjunction with Lebesque's term-by-term integration theorem to show that two functions f and g that have the same Fourier expansion must be identical everywhere with the possible exception of a set of measure zero. Now assume that

$$f(x|\theta) = \sum_{k=-m}^{m} B_k \exp(2\pi ikx),, \tag{9.1.1}$$

where, here, the Fourier coefficients $\{B_k\}$ are the elements of the parameter row vector $\theta = (B_{-m}, B_{-m+1}, \ldots, B_{-1}, B_0, B_1, \ldots, B_{m-1}, B_m)$. For any index k

$$\partial \log f(x|\theta)/\partial B_k = \frac{\exp(2\pi ikx)}{\sum_{s=-m}^{m} B_s \exp(2\pi isx)}. \tag{9.1.2}$$

Partial derivative (9.1.2) is a continuous function of x and $\{B_k\}$. Thus Leibnitz' rule, Taylor, (1955, page 525) can be applied to show that likelihood function (8.2.3) of the previous chapter attains its maximum value for some sequence $\{B_s^*\}$ such that

$$\sum_{v=-m}^{m} \int_0^1 \left[\hat{B}_v \exp(2\pi i(k+v)x) \Big/ \sum_{s=-m}^{m} B_s^* exp(2\pi isx) \right] dx=0 \tag{9.1.3}$$

for $k=0, \pm 1, \pm 2, \ldots, \pm(m-1), \pm m$. For any given k, (9.1.3) is a finite sum. Thus, (9.1.3) yields the system of equations

$$\int_0^1 \left[\frac{\sum_{s=-m}^{m} \hat{B}_s \exp(2\pi isx)}{\sum_{s=-m}^{m} B_s^* \exp(2\pi isx)} \right] \exp(2\pi ikx) dx=0 \tag{9.1.4}$$

for $k=0,\pm 1,\pm 2,\ldots,\pm(m-1),\pm m$. Now suppose K^{-1} is defined as the special case of the integrand of (9.1.4) when $k=0$. The function $y(x)$ can then be defined as

$$y(x) \equiv \left[\sum_{s=-m}^{m} \hat{B}_s \exp(2\pi isx)\right] \Big/ \left[\sum_{s=-m}^{m} B_s^* \exp(2\pi isx)\right] - K^{-1}. \quad (9.1.5)$$

Equations (9.1.4) imply that as $m \to \infty$ all the Fourier coefficients of $y(x)$ will be zero. Hence, when g satisfies the conditions of the Edwards (1967) or Sz-Nagy (1965), uniqueness theorem, almost everywhere, (9.1.5) will be identically equal to zero. Consequently,

$$\sum_{s=-m}^{m} B_s^* \exp(2\pi isx) = K \sum_{s=-m}^{m} \hat{B}_s \exp(2\pi isx)$$

almost everywhere. Defining

$$f^*(x) = \sum_{s=-m}^{m} B_s^* \exp(2\pi isx),$$

the constraint $\int_0^1 f^*(x)dx=1$, implies $B_0^*=1$. Therefore, the uniqueness theorem implies that $B_s^*=\hat{B}_s$ for $s=1,2,\ldots,m$. Hence, substituting each ML estimator for the respective Fourier coefficient of (9.1.1) yields the raw estimator

$$f(x|\theta) = \sum_{k=-m}^{m} \hat{B}_k \exp(2\pi ikx) \quad (9.1.6)$$

9.2 Survival curve applications: The stationary case

We will now consider both univariate survival variate curve estimation and the estimation of regression curves where the dependent variate is a measure of survival. Generally speaking, survival analysis applications can be divided into two categories: the stationary and nonstationary cases. In this section we will consider the stationary case. In the next section we will consider several different variations of the nonstationary case where it is not assumed that the survival time and the date of diagnosis are independent variates. In Section 9.3 we will consider how the $R(x)$ and $R(x)/R'(x)$ extended model parsing process can be generalized to deal with nonstationary survival analysis and other bivariate applications.

The reasoning that leads to (9.1.6) can be used to estimate a survival curve S from a sample consisting of both complete survival times and incomplete measurements where survival is only known up to a date of study completion. This composite structure of survival data obtained from most clinical trials is what motivates much of life table methodology and the Kaplan-Meier (1958) product limit limiting case of the life table. By using reasoning similar to the above in conjunction with the Bartholomew (1957) likelihood function which underlies most survival curve analysis, a simple closed-form modular expression for the survival MLE can be obtained in terms of the estimated density of the complete subsurvival and the incomplete subsurvival functions. Here the term modular refers to the same concept that underlies Peterson's (1977) use of the prefix "sub" of the term "subsurvival function." In terms of the methodology discussed in Chapters 6 and 7 modularity distinguishes *extended models* from models, such as the normal and extreme value densities, which are expressed as a specified elementary functions.

Subsurvival functions are the modular components of the extended model of the overall survival function that can be estimated separately, just as the individual components f_1 and f_2 of the mixture $f(x|p,\mu_1,\mu_2,\sigma) = \{pf_1([x-\mu_1]/\sigma) + (1-p)f_2([x-\mu_2]/\sigma)\}/\sigma$ can be separately estimated by the techniques described in Chapter 5 or the components G and v of $G\{(\text{v}[x-\mu_1]-\mu_2)/\sigma\}$ separately selected by the techniques described in Chapters 6 and 7. Note however, for the case of the subsurvival approach, unlike the mixture extended model, it is known a priori which data point corresponds to which subsurvival function. Thus, in subsurvival applications, the selection or separation of two component curves is not a point at issue. Instead, emphasis is placed on the reconstitution of a whole using information concerning its parts.

By using the reasoning that yielded (9.1.6), Tarter, Freeman and Polissar (1990) showed that the modular extended model approach can substantially improve the estimation of long-term survival rates. Simulation studies using Weibull and lognormal data show that even in the case found to be least favorable, the modular method has less than one seventh the prediction error of all conventional life table, LT, or Kaplan-Meier, KM, estimators, even when the LT and KM techniques are optimized for the purpose of long-term survival estimation. In addition to conventional survival applications, the modular ML procedure can be used to estimate the probability of being disease-free at different ages and following different exposures to possibly harmful environmental contaminants. It is also shown that the modular approach is particularly useful in situations where the effects of a confounding, nuisance, or effect-modifying variable cannot be confidently expressed as an elementary function.

To illustrate how modular nonparametric subsurvival estimation works, consider the following example: From 1940 to 1974, $n=780$ patients were diagnosed at the University of California, San Francisco, as having breast cancer with regional spread. The median post diagnosis survival time of those patients who died before 1974, and thus yielded complete measurements, was 72 months. Figure 9.1 displays estimates of the subsurvival curves corresponding to the complete and incomplete observations, where the former were obtained from those patients who died before 1974 and, for those patients who lived past the 1974 study

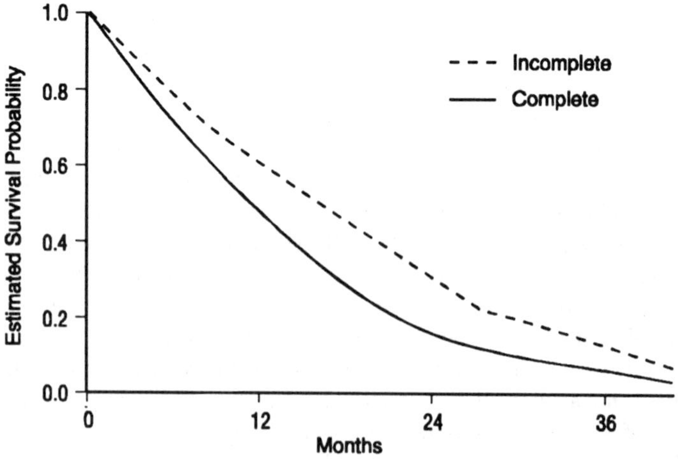

Figure 9.1. *Estimates of subsurvival functions corresponding to the complete and incomplete measurements from a study of 780 breast cancer patients.*

termination date, the latter equals the time between date of diagnosis and the last date, t_m=December 31, 1974, at which death could have occurred and this event also be recorded for later use.

To obtain the curves shown below, individually optimized truncated Fourier expansions were constructed from the samples of completes and incompletes, respectively, using the technique described in Section 4.7. These individually optimized estimators were then combined to estimate the survival distribution of the population. The smoothness inherent to generalized forms of representation (as well as the separate optimization of the two constituent subsurvival estimators) substantially improves the stability of the estimated survival distribution (especially in the tails). The inner curve shown in Figure 9.1 corresponds to the complete measurements. As is typical, the estimated subsurvival function associated with the subpopulation of individuals who were alive at t_m was substantially different from that corresponding to the complete subpopulation.

There is a basic difference between the extended model and alternative methodologies, such as the clinical life table (LT) and Kaplan Meier (KM) product limit procedure. For the former, curve estimation techniques are used to estimate *both* of the two distinct subsurvival functions. When the number of incomplete measurements is either greater than, or not substantially smaller than, the number of complete measurements, two curves similar to those shown in Figure 9.1 are estimated. Otherwise, two curves (one of which is associated with the *combined* complete and incomplete subpopulations) are constructed. Thus, it is the similarity of approach applied to the constituent parts of an extended model that differentiates the extended model approach from the LT-KM approach to survival curve estimation.

The capacity to optimize separately each of the two constituent subsurvival curves and then form a composite estimator is greatly enhanced by the application of likelihood methodology. The following relationship is shown in Tarter (1979a): Suppose $1-\hat{F}_1$ and $1-\hat{F}_2$ represent ML estimators of the m-term truncated Fourier expansion model for the complete and incomplete subpopulation survival curves, respectively, and $\hat{f}_1$ is the derivative of $\hat{F}_1$. If $\hat{S}(t)$ represents the function that maximizes the Bartholomew (1957) likelihood function then

$$\log\hat{S}(t) = \int_0^1 \frac{\hat{f}_1(x)}{\hat{F}_1(x)+(n_2/n_1)\hat{F}_2(x)-(n_2+n_1)/n_1}dx, \qquad (9.2.1)$$

where n_1 and n_2 represent the number of iid complete and incomplete observations, respectively.

The dashed curve shown in Figure 9.2 is a modular estimator based on (9.2.1), while the dotted curve depicts a KM estimate. Both of these estimators are based on the data used to obtain the subsurvival functions $1-\hat{F}_1$ and $1-\hat{F}_2$ shown in Figure 9.1. Figure 9.3 is identical to Figure 9.2 except that it shows a life table in place of the KM estimator shown in Figure 9.1. The degree of abridgment, i.e., the class interval length, of this life table was determined by dividing the range of the combined sample of all survival measurements by 12. (Note that the KM can be conceptualized as the limiting case of the life table where the class interval length approaches zero.) Figure 9.4 shows a close-up image of the curves shown in Figure 9.3 for the survival period of 50–60 months.

The near identity of the curves shown in Figure 9.4 gives the false impression that there is little difference between modular extended model (MEM) methods and conventional KM-LT methodology. While it is true that for survival times which are of short and moderate duration there are indeed few differences (at least in the univariate case), there is a MEM approach which is of considerable practical utility. By using the MISE weighting techniques described in Chapter 4, Tarter, Freeman and Pollisar (1990) showed that methods for estimating long-term sur-

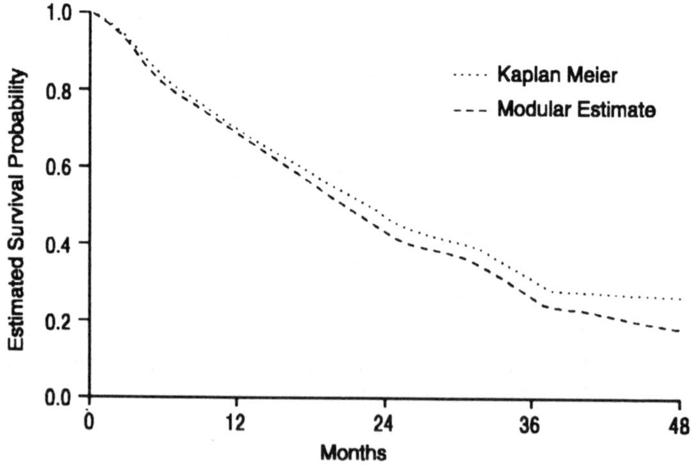

Figure 9.2. *Comparison of Kaplan Meier estimate and the modular survival estimate constructed from the two subsurvival functions shown in Figure 9.1.*

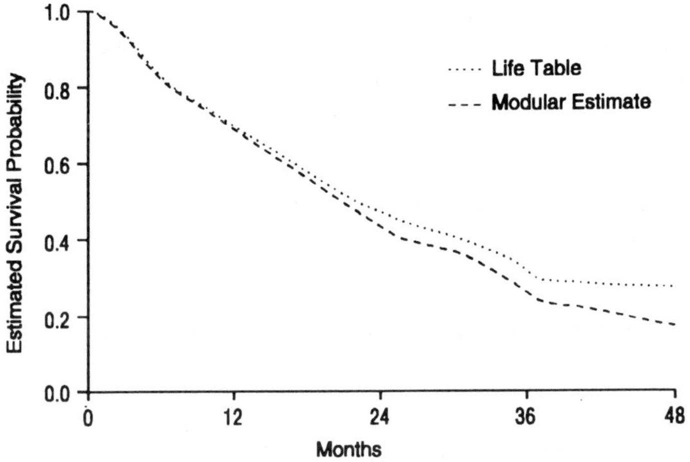

Figure 9.3. *Comparison of life table estimate and the modular survival estimate constructed from the two subsurvival functions shown in Figure 9.1.*

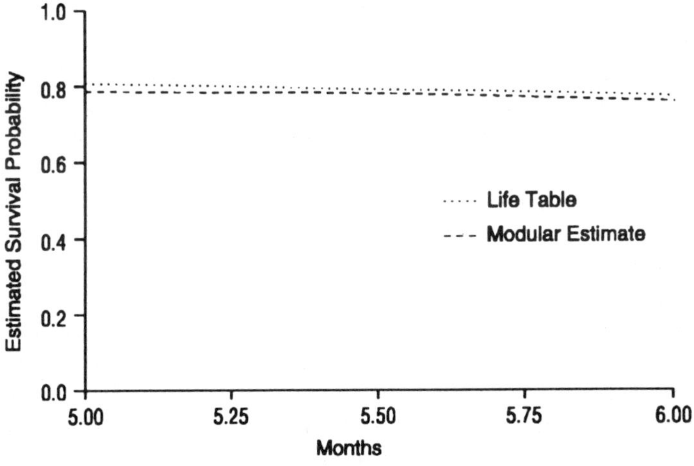

Figure 9.4. *Close-up comparison of the two estimates shown in Figure 9.1 over the survival period of 50–60 months.*

vival could beimproved substantially. Simulation studies using Weibull and lognormal data showed that even in the case found to be least favorable (out of 16 sets of trials with different distributions and different parameters) the new method has less than one-seventh the prediction error of all conventional LT-KM estimators, even though the LT estimators are optimized for the purpose of long-term survival estimation, in other words, optimized using prior knowledge of the true survival function, knowledge not assumed as part of the construction of the MEM estimators.

9.3 Survival curve applications: The nonstationary case

Let the letter t represent a value attained by the survival time variate T corresponding to an individual who has been diagnosed at date $X=x$. Let ω represent the earliest date at which study subjects could have been diagnosed and Ω represent the latest possible date of diagnosis. In other words, the study's data collection interval is $[\omega,\Omega]$. It is therefore obvious that for an individual to yield one of the complete measurements described in the last section, $x+t\leq\Omega$. Suppose that $f(t|x)$ is the probability that an individual who is diagnosed as having a certain disease at date $X=x$ will survive t time units after x. In some applications $f(t|x)$ could represent the chance that a machine which is put into service at date x will fail at date $x+t$. It is important to distinguish the function f from the hazard function h and instantaneous specific death rate q. Both h and q are not only conditioned on date $X=x$ but are also conditioned on the event that the individual in question has survived up to, but not beyond, date $x+t$. Here $h(t|x)=f(t|x)/S(t|x)$, where survival curve

$$S(t|x) = 1-F(t|x) \text{ and } F(t|x) = \int_0^t f(y|x)dy. \tag{9.3.1}$$

As introduced in the previous section, the actual sample of data used to estimate $h(t|x)$, $f(t|x)$ and $S(t|x)$ is almost invariably comprised of two portions; the complete measurements whose values satisfy the inequality $x+t\leq\Omega$, and the incomplete measurements where $x+t>\Omega$, whose individual member is given the value $\Omega-x$, usually referred to as either a known survival time or a period of observation. In practice it is important to distinguish varieties of censored observations of which incomplete measurements are a subset. Strictly speaking, censoring of the incomplete variety occurs when an individual is known to be alive at Ω.

Owing to data precision considerations, an individual could be said to die at Ω exactly. However, one of the advantages of the subsurvival approach is that the Sheppard's correction type procedures can be as easily applied to survival analysis applications as it can to applications that do not involve modular estimates of two subpopulation densities. Hence, we will assume below that such correction methods have been applied.'

Besides incompleteness, censoring can be attributable to the withdrawal of subjects prior to Ω. Patients can, of course, be lost to follow-up for a variety of reasons, among which is change of address to a location that cannot be determined by those conducting the study. In the remainder of this chapter the subscript w will be used to represent withdrawal and will have no connection with the mean integrated square error weight.

Below, $f(t|x)$ will represent the distribution density for the limiting case where $\Omega \rightarrow \infty$. We will call the hypothesis or assumption that $f(t|x)$ is functionally independent of x, $H^{(1)}$: $f(t|x)=f(t)$, the stationarity hypothesis. As far as the withdrawals are concerned, the following three assumptions or hypotheses can be made concerning $f_w(t|x)$, the conditional density of the withdrawals' distribution, $H^{(2)}$: $f_w(t|x)=f(t|x)$, representative withdrawal, $H^{(3)}$: $f_w(t|x)=f_w(t)$, stationary withdrawal, and $H^{(4)}$: $f_w(t|x)=f(t)$, representative-stationary withdrawal. $H^{(2)}$ applies when the events that determine withdrawal are unrelated to either date of diagnosis or extent of illness. If patients who are diagnosed when they are already terminally ill tend to be lost because of travel to some distant location in search of some miracle cure, and if this tendency to lose information is unrelated to date of diagnosis, $H^{(3)}$ would be an appropriate choice. Because it is a composite of $H^{(1)}$ and $H^{(2)}$, $H^{(4)}$ is the most unlikely choice in terms of validity, but the most desirable choice in terms of mathematical tractability.

In some circumstances incompletes and withdrawals can be combined prior to the implementation of the curve estimation procedures described in this and the previous section. In the remainder of this chapter, we will assume that $H^{(1)}$ is not valid, but that no distinction need be made between withdrawals and incompletes, which will be classed together under the single term, incomplete. To check whether incompletes and withdrawals should be pooled, the survival model and parameter selection methodology described in the next section, the inferential procedures described in Chapter 10, or any one of many alternative means, can be utilized.

Suppose a sample of survival time and/or date of diagnosis measurements is divided into two components as follows: $\{(T_j,X_j)\}$, $j=1,\ldots,n_1$

consist of complete measurements and $\{X_j'\}$, $j=1,\ldots,n_2$ consist of the dates of diagnosis of incomplete measurements, where the jth member of $\{X_j'\}$ is an individual whose unknown survival time is T_j'. Tarter and Freeman (1983) showed that if $\{(T_j,X_j)\}$, $j=1,\ldots,n_1$ and $\{T_j',X_j'\}$, $j=1,\ldots,n_2$ are iid with densities that have a uniformly convergent bivariate Fourier expansions, then the ML estimate of this common density can be expressed as $(n_1/n)\hat{f}(t,x)$ within the part of the x-y plane where $x+t\leq\Omega$. Here $\hat{f}(t,x)$ is some bivariate Fourier series estimator computed from the subsample $\{(T_j,X_j)\}$, $j=1,\ldots,n_1$.

No estimate of the conditional $f(t|x)$ can be expected to be accurate over that part of the x,y plane where $x+t>\Omega$. Nevertheless, we now see how $(n_1/n)\hat{f}(t,x)$ can be applied to provide useful information concerning survival variates in the nonstationary case. Suppose the density of times of diagnosis can be estimated from the pooled sample $[\{X_j\}, j=1,\ldots,n_1, \{X_j'\}, j=1,\ldots,n_2]$ as $\hat{f}^{(CI)}(x)$. Then $\hat{f}(t,x)/\hat{f}^{(CI)}(x)$ is an estimate of the conditional density $f^{(C)}(t|x)$ of survival time t, given date of diagnosis x, as well as given that a subject will yield a complete observation, in other words, that the values of the observations, x and y, satisfy the inequality $x+t<\Omega$.

What one would like to examine in most circumstances is not $f^{(C)}(t|x)$ but the mixture

$$I_{[x+t\leq\Omega]}(t,x)p^{(C)}(x)f^{(C)}(t|x)+I_{[x+t>\Omega]}(t,x)p^{(I)}(x)f^{(I)}(t|x) \qquad (9.3.2)$$

which characterizes the combined distribution of both complete and incomplete measurements. Here $p^{(C)}(x)$ and $p^{(I)}(x)$ are, respectively, the probability given x that an observation will be complete, and $1-p^{(C)}(x)$. Because $f^{(C)}$ equals zero when $x+t>\Omega$, the use of the term $I_{[x+t\leq\Omega]}(t,x)$ is actually not needed in (9.3.2). However, it is included to help interpret the method described below. Also, without additional information it is impossible to find a reasonable estimator of $I_{[x+t>\Omega]}(t,x)p^{(I)}(x)f^{(I)}(t|x)$.

We will see that the $R(x)$ methodology described in the next section can be applied, even if the pdf and cdf components of $R(x)$ are only known in some subinterval of a variates support region. In this way nonstationary survival processes can be studied in terms of the properties of $p^{(C)}(x)f^{(C)}(t|x)$ estimated over the region $x+t\leq\Omega$. As described in Tarter and Freeman (1983), Bayes' theorem can be used to obtain the estimator $(n_1/n)\hat{f}(x|c)/\hat{f}^{(CI)}(x)$ of $p^{(C)}(x)$. Obviously, estimates of $p^{(C)}(x)f^{(C)}(t|x)$ will be most reliable for small values of x and t. For these values, we will now proceed to discuss how these and other bivariate estimators can be used to select an appropriate model G and variate transformation v. These methods will also allow the structure of $f(t|x)$ to be described in terms of particular parameters of the bivariate extended models, which describe the distribution of the survival and time of diagnosis variates.

9.4 Bivariate model and transformation selection

The geometric properties of the R function described in Chapters 6 and 7 can be used in the bivariate case to expedite greatly the selection of bivariate forms of parametric representation in nonstationary survival and other bivariate data applications. Consider the four functions, $f_s(y|x_s)$ and $F_s(y|x_s)$, $s=1,2$, where $x_1 \neq x_2$, and where $f_s(y|x_s)$ and $F_s(y|x_s)$ are the two densities and two cumulatives of two conditional distributions indexed by $s=1$ and $s=2$, respectively. Then these four functions can be assembled to form the two functions $R_s(y|x_s)$, $s=1,2$, where R_s is defined in Chapters 6 and 7.

Let $\mu_{(1,s)}=\tau_s$, $\mu_{(2,s)}$, and σ_s, $s=1,2$, represent the pre and post-transformation location parameters and the scale parameters of two logmodels $G\{(\log[y-\tau_s]-\mu_2^{(s)})/\sigma_s\}$, $s=1,2$. Note, below we will use the symbol τ_s, in place of the $\mu_{(1,s)}$, because the former symbol requires only a single subscript. It was shown by Tarter and Freeman (1987) that when $R_s(y|x_s)$, $s=1,2$, are determined using the correct choice of the model G, then whenever $\sigma_1=\sigma_2$, any two $R_s(y|x_s)$, $s=1,2$, will be a pair of parallel lines. In addition, all $R_s(y|x_s)$ constructed from distributions for which it is valid to assume a common pretransformation location parameter τ, must be lines that pass through a single point. We will also show that these curves can be estimated for the purpose of studying survival distributions in the nonstationary case.

Figures 9.5 and 9.6 illustrate how the above result can be applied. Here, to show that is was typically possible to obtain informative displays from bivariate samples of size $n=200$, Figure 9.5 displays estimates of $R_s(y|x_s)$, $s=1,\ldots,7$, and Figure 9.6 shows estimates of $R_s(y|x_s)$, $s=1,\ldots,5$ for $n=200$ samples of iid data where G was the standard normal cdf. For Figure 9.5, $\sigma_s=0.7$, for all indices $s=1,\ldots,7$; while τ_s, $s=1,\ldots,7$ were chosen at equally spaced, and therefore different, values. Hence, the curves displayed estimate a system of parallel lines.

Figure 9.6 illustrates the opposite application of bivariate R. Here

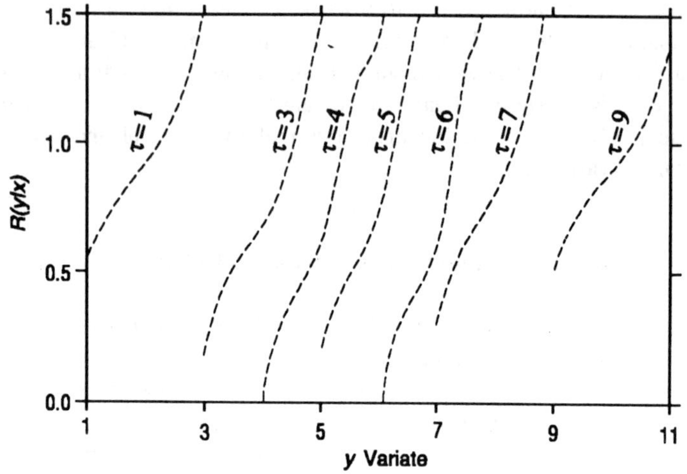

Figure 9.5. *Illustration of R(y|x). Here is a lognormal distribution where the pretransformation parameter, τ, changes with x while the standard deviation (σ=0.7) remains fixed. The numbers next to each line are the true pretransformation parameters. Sample size was 200.*

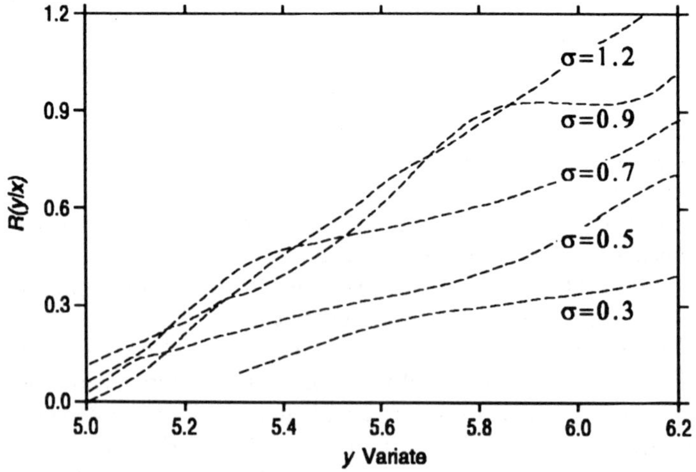

Figure 9.6. *R(y|x) is demonstrated where f(y|x) is a lognormal distribution where the standard deviation σ changes with x while the pretransformation parameter (τ=5) remains fixed. The number next to each line is the true standard deviation of the sample. Sample size was 200.*

for the same G and n, $\tau_s=5$, for all indices $s=1,\ldots,5$; while σ_s, $s=1,\ldots,5$ were chosen to be an equally spaced sequence of different values. In this example, as expected under the assumption that the pretransformation parameter represented by the letter τ, is constant, symbolically $\tau(x)\equiv C$, the R curves appear to converge to a single point. Samples of size $n=200$ were used to test whether the assumption of homoscedasticity could be checked. For this purpose, and to check the assumption of equal τ, the experiments described in Figures 9.5 and 9.6 were each repeated 100 times. Figures 9.7 and 9.8 are box and whisker plots, which correspond to Figures 9.5 and 9.6, but where box boundaries were set at the estimated first and third quartiles and whiskers extend from the 2.5th to the 97.5th percentiles of graphed values.

To study the case where a normal rather than a lognormal bivariate data set is encountered, the same experiment was repeated for normal data where the standard deviations of normal conditionals were functionally related to the x variate. As illustrated in Figure 9.9, the estimated curves were fairly close to straight horizontal lines over the intervals depicted.

The methods described above can be used to check on a very general

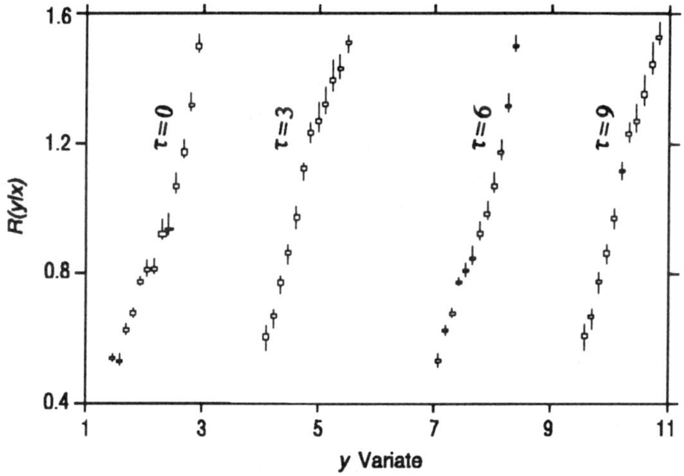

Figure 9.7. *The experiment described by Figure 9.5 was repeated 100 times to produce the average depicted here. The numbers next to each line are the true pretransformation parameters of each sample. The average estimated parameters for each sample were (from left to right)—0.1, 2.4, 5.9, and 8.6.*

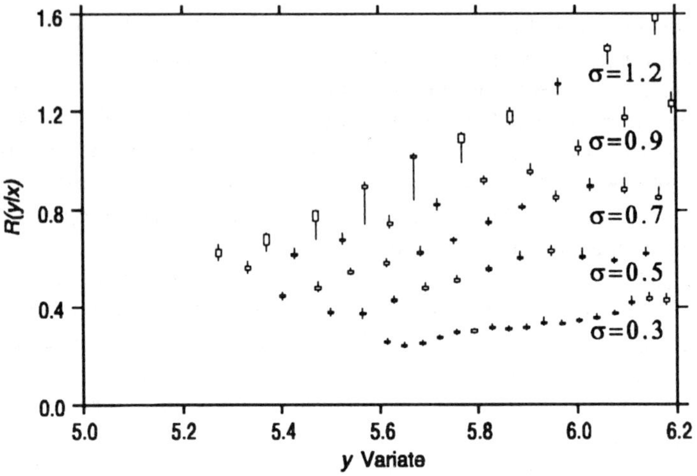

Figure 9.8. *The experiment described by Figure 9.6 was repeated 100 times to produce the averages depicted here. The numbers next to each line are the true standard deviations of each sample. The average estimated standard deviations of each sample were (from top to bottom) 0.27, 0.54, 0.67, 0.76, and 1.07.*

class of assumptions. They can be applied, as they were in Chapter 7, to the fractional power transformation, $v(z)=[z-\tau]^{[1/(1+K)]}$, special case of the v-model within $G([v(y)-\mu_2(x)]/\sigma)$. For this v-model, the function of y formed by dividing the function R by its derivative R' will be identically equal to the linear function $[(1-K)/K](y-\tau)$. Thus, if the curve $[R/R'](y|x_s) \equiv R(y|x_s)/R'(y|x_s)$ is used in place of the curve $R_s(y|x_s)$ for $s=1,\ldots,7$ described above, a sequence of parallel lines will be generated in all cases where a common value of K within the fractional power transformation implies that a single fractional power transformation will transform all data points to variates with cdf $G\{[y-\mu(x)]/\sigma(x)\}$. The actual common value of K, and the value of $\tau(x)$ for any conditionals $f(y|x)$ and $F(y|x)$ can be determined from the slope and intercept of the curve R/R'. Most importantly, with the one proviso that $\tau(x)$ is not a horizontal line, the conditionals $[R/R'](y|x_s)$, $s=1,\ldots,7$ will be a sequence of parallel lines no matter what form is assumed by the three curves $\mu(x)$, $\sigma(x)$ and $\tau(x)$. When $\tau(x)$ can be validly assumed to be a horizontal line, all $[R/R'](y|x_s)$ curves will meet at a single point.

When $[R/R'](y|x_s)$ curves are not parallel lines, but instead, when

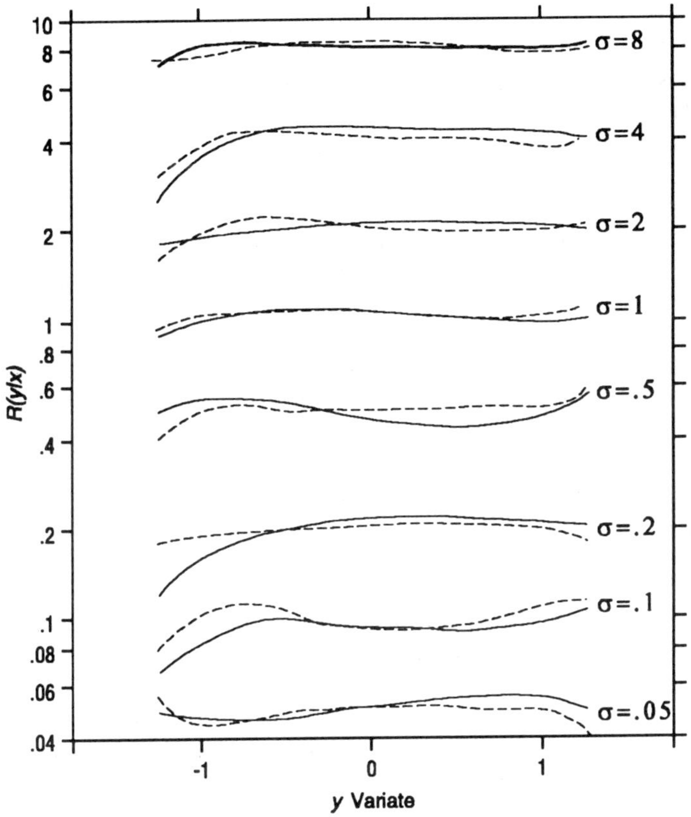

Figure 9.9. *Illustration of* $R(y|x)$ *where* $f(y|x)$ *is a normal distribution where the standard deviation,* σ, *is functionally related to x. Here* $\sigma(x_j) \equiv x_j$, *where* $x_j, j=1,\ldots,8$ *are the numbers displayed along the right side of the graph. The solid lines correspond to cases where* $\mu=\sigma(x_j)$ *while the broken lines correspond to* $\mu=10\sigma(x_j)$. *Sample size was 200 for each trial.*

they meet at a single point, this configuration implies that the curve $K(x)$ cannot be a horizontal line. In other words, more than a single fractional power transformation will be needed to transform bivariate data. However, this configuration also implies that no matter what the functions $K(x)$, $\mu_2(x)$, and $\sigma(x)$ are, the curve $\tau(x)$ will be a horizontal line. In other words, a single pretransformation location parameter (threshold parameter) will suffice for all x values.

In the bivariate case there are so many different ways to use $[R/R'](y|x_s)$ and $R(y|x_s)$ that it is useful to describe a systematic strategy by which nonstationary survival and other forms of data can be studied. The display of $[R/R'](y|x_s)$, $s=1,\ldots,c$, curves can be a valuable first step. The number of curves, c, graphed and the spacing of these curves depends on the amount and quality of data. When small to moderate samples, $n=100$ to 250 are used, it will be best to choose values of x_s, $s=1,\ldots,c$, which are close to the mean of the x variate.

If, for a given choice of G, $[R/R'](y|x_s)$, $s=1,\ldots,c$, contains few or no straight lines, mixture decomposition approaches, or alternative choices of G, should be considered. That displays based on $[R/R'](y|x_s)$ yield lines, but lines that are neither parallel nor coincident at a single point, suggests that fractional power transforms are applicable, but that neither $\tau(x)$ nor $K(x)$ are horizontal lines. In other words, τ and K are nonconstant functions of x. Consequently, at each value of x_s a particular $K(x)$ of the fractional power transformation and a particular $\tau(x)$ must be selected by using the relationship between K and τ, and the slope and intercept of the line $[R/R'](y|x_s)$.

The observation that $[R/R'](y|x_s)$, $s=1,\ldots,c$, are linear, but that lines are either parallel or coincident at a single point, but not both, suggests that fractional power transforms are applicable, but that either the same K can be applied for all x, the case of parallel lines, or that the same τ can be applied for all x. In either case, an average of the common parameter K or τ can be used to estimate the common parameter value. Values of the other parameter can then be estimated separately for each line.

Once values of K and τ are obtained, the sequence of c different $R(y|x_s)$ lines can be applied to the estimation of the parameters σ_s, $s=1,\ldots,c$, of the fractional case, by raising each estimate of $R(y|x_s)$ to the $[(1+K)/K]$ power, then applying the relationship $[R(y|x_s)]^{[(1+K)/K]}=\sigma(y-\tau)$, $s=1,\ldots,c$; which also will provides a check on the value or values of $\tau(x)$. In some situations where a large value of K is determined to be the most appropriate choice for this parameter, it will be more convenient to apply the logarithmic transformation than to apply the fractional power transformation with this value of K. When $\sigma<1$, we have found that in most circumstances a fractional power transformation where $K>4$ and the logarithmic transformation will yield nearly identical results.

When $[R/R'](y|x_s)$ methodology yields diagonal lines, all with a slope of 45°, and when $R(y|x_s)$ are not horizontal lines, this suggests that the logarithmic limiting $K \rightarrow \infty$ case should be applied in place of a fractional power transformation with a finite K. In this situation, the steps illustrated by in Figures 9.5 through 9.9 can be taken to first ascertain which, if any, logmodel parameters are functionally related to x.

If one is extraordinarily lucky in survival curve applications, where G is chosen to be the standard extreme value cdf, and other applications where G is a standard normal's, or some other model's cdf, $R(y|x_s)$ plots will be found to be congruent horizontal lines. In this case, conventional model based parametric methods can be safely applied.

9.5 Likelihood ratio tests

Once clues and conjectures concerning the structure of the survival density or hazard function $f(x)/S(x)$ have been brought to light by the methods described in the previous sections, hypothesis testing procedures can then be implemented to validate these conjectures. While general testing procedures will be described in the next chapter, we will end this chapter by briefly outlining how the new likelihood estimation techniques can be adapted to form likelihood ratio tests.

Consider the closed-form nonparametric maximum likelihood curve estimator $\hat{S}$ of the survival curve S, where $\hat{S}$ is expressed in terms of separate curve estimates obtained from samples of randomly censored and uncensored times to failure, that is, subsurvival populations. The log likelihood (logL) function constructed from such data equals

$$\sum_{j=1}^{u} \log[-S'(Y_j)] + \sum_{j=1}^{v} \log[S(Z_j)],$$

where $\{Y_j\}$, $j=1,..,u$ are complete, and $\{Z_j\}$, $j=1,..,v$ are the incomplete observations.

To devise a likelihood ratio test of the null hypothesis that two survival populations are identically distributed, the Neyman Poisson procedure described in Section 8.6 can be generalized yield the expression

$$\log L_1 = \sum_{j=1}^{u} \log[-\hat{S}'(Y_j)] + \sum_{j=1}^{v} \log[\hat{S}(Z_j)],$$

where $\hat{S}$ is a survival curve estimator constructed by using the combined data from the two populations. Similarly, data from the first population can be used to estimate $\log L_2$ and then, separately, data from the second population can be used to estimate $\log L_3$. A likelihood ratio test of the null hypothesis of population identity can then be based on the test statistic $\log L_1 - \log L_2 - \log L_3$.

By taking the difference between the log likelihood function described in Section 9.2 and the log likelihood function described in Section 9.3, a likelihood ratio test of the null hypothesis of stationarity against the alternative hypothesis of non-stationarity, similar to the test based on $\log L_1 - \log L_2 - \log L_3$, can be devised. With regard to the number of degrees of freedom of the chi-square distribution that would yield the best chi-square approximation under the null hypothesis of nonstationarity, very little is known. However, the number of Fourier series coefficient estimators used in any application is at least denumerable; unlike the number of support points of the continuous variate generalization of the Neyman-Poisson discrete model procedure. Hence, Neyman's procedure can be extended to continuous variate likelihood ratio applications based on Fourier representation, which include tests of survival variate stationarity.

The above likelihood ratio procedure is discussed in this chapter primarily to illustrate the applicability of series-based likelihood methodology. However, before changing emphasis to questions concerning the power curves and other criteria by which inferential methods are judged, the following application involving the transformation v, discussed in this chapter and in chapters 6 and 7, will be described.

Suppose v_1 and v_2 are two candidate transformations to normality as defined in Chapter 7. Based on an iid sample $\{X_j\}$, $j=1,..,n$ the difference of the two log-likelihood functions—the first corresponding to v_1 and the second corresponding to v_2—equals

$$-n\log[s^{(1)}/s^{(2)}] + \sum_{j=1}^{n} \log[v_1'(X_j)/v_2'(X_j)]$$

where the squares of $s^{(1)}$ and $s^{(2)}$ are the ML estimates of the scale parameters of the transformed variates. It was shown in Chapter 7 that the normal density can be treated as a limiting case of a properly scaled lognormal density whose parameter $\mu_1 \to \infty$. Since the normal distribution is a special limiting case of the lognormal distribution, it follows that when $v_1'(X_j)=1$, $v_2'(z)=\log(z+\hat{C})$, and $\hat{C}$ is the ML estimator of μ_1,

$$\log L = -n\log[s^{(1)}/s^{(2)}] + \sum_{j=1}^{n} \log[v_1'(X_j)/v_2'(X_j)]$$

is a likelihood ratio test. Relationships described in Chapter 13 of Wilks (1962) indicate that $-2\log L$ converges in distribution to the single-degree-of-freedom chi-square distribution, which Wilks (1962) designates as C(1).

It is possible to expedite the computation of logL. The only computer-time-intensive step needed to compute logL is the determination of μ_1. A computationally and data-usage efficient estimator of this constant can be obtained by using numerical differentiation methods to evaluate the tangent of $R(x)$ at $\hat{x}$, where $\hat{x}$ is the estimated mode of f. The estimator can either be used to approximate the true MLE of μ_1 or can be used as an initial value for the iterative computational routine for the ML estimation of μ_1 and other model parameters. Its efficiency stems from the center-weighting characteristics of MISE $J(\hat{f},f,1)$ that was discussed in Chapter 4.

CHAPTER 10

Nonparametric curve estimation and inference

In this chapter we will take the liberty to speculate on one area that, of necessity, is presented with less detail than topics discussed in previous chapters. The reason for this lack of detail is that the two-valued loss function which underlies frequentist forms of inferential statistics is fundamentally different from the MISE metric upon which much of the previously described methodology was based. Pioneering methodology devised by Watson (1967), and all of the other inferential methods described below, have some connection with MISE. Nevertheless, to be truly comparable to conventional inferential procedures, the performance of generalized methods should primarily be gauged in terms of power, $1-\beta$. Power is, of course, merely the chance that by using a particular test, one can say what one wishes to say, when one should say it. One either says something correctly, or incorrectly, and in the latter case, one incurs some constant loss. Therefore, unlike the case of the MISE metric, there is no continuum between small, intermediate and large losses.

As far as we are aware, with the exception of the Schweder-Hodges-Lehmann (SHL) asymptotic nonparametric procedure briefly described in Section 10.4, there are at present no generalized methods that are specifically designed to have optimal power characteristics, rather than designed in accordance with some curve-fitting criterion such as MISE or penalized likelihood. This lack is primarily due to computational considerations. For example, the power curves shown below in Figure 10.2 are used to assess the performance of the hybrid test based on a series optimized with respect to MISE. It is feasible, but would require considerable conceptual and computational effort, to use the evaluation of one of these power curves to replace the MISE as a KT selection criterion.

Despite the many points of departure between inferential and exploratory applications of generalized forms of statistical representation there are also some commonalities and mutualities. One example is the analog of least-squares-fitting criteria, which, by means Fourier series-based procedures, can estimate the same location parameter that can be estimated and tested by classical procedures. Similarly, in the previous chapter new generalized likelihood methodology was shown to be applicable to the same survival analysis problem commonly approached using the *nonparametric* life table and its Kaplan-Meier product limit limiting case. As briefly outlined at the end of the last chapter, this methodology can be used to construct likelihood ratio tests of survival curve identity.

Much of this book concerns complexities of mixture decomposition, a topic which presently is the subject of very little literature that can be called "inferential." According to P.D.M. Macdonald (1975): "Over the past 80 years a bewildering variety of methods have been proposed for estimating the parameters of distribution mixtures." Hence, is not surprising that the methods referred to by Macdonald include some which are based on curve estimation techniques like the lambda method detailed in Chapter 5. In view of the difficulty of mixture problems, it is also not surprising that decomposition techniques do not as yet belong to the mainstream of statistical practice. For example, although cluster analysis routines are occasionally included as part of off-the-shelf statistical programs, neither the semiparametric lambda approach nor any parametric Pearson-moment-based mixture decomposition approach is currently available in any of these packages.

It also should be pointed out that although the procedures described in Chapter 5 can often facilitate the decomposition of a mixture into subcomponents, help identify some candidate model for the subcomponent and then find a ML-type estimator of the subcomponent's parameters, there is as yet no published research that deals with hypothesis tests involving the values of these parameters. Inferential procedures described in the next section are based on information accessed in the frequency domain. Taking this new viewpoint increases the practicality of inference concerning the structure of two mixtures. For example, suppose two independent samples are used to study two populations, each a mixture containing two distinct subpopulations. Suppose further that prior knowledge indicates that if there is a difference between the populations, this distinctiveness can only be due to a difference between the upper mixture components. A test of shift between the upper mixture components can be implemented by first using techniques described in

Chapter 5 to estimate the Fourier coefficients of each population's upper component and then using these estimates to perform the shift test described in the next section.

In one respect, the methods described in this chapter are the opposite of mixture decomposition procedures. To date, the only places in the statistical literature where hypothesis testing and generalized mixture decomposition intersect are the bump-hunting approach of Good and Gaskins (1980) and the mode-based methods described by Izenman and Sommer (1988). With these exceptions, inferential procedures have dealt with problems simpler than mixture decomposition.

The broad area of confirmatory statistics already has an extremely large literature. Unlike decomposition, it is already well-represented within standard computer packages. Also, unlike mixture studies, whose literature is at present heavily weighted toward the theoretical side, both frequentist and Bayesian approaches to inference are commonly used tools of statistical practice. In view of the power and scope of these already highly perfected tools, curve estimation has not yet had a substantial impact on the fields of nonparametric, Bayesian and classical inference. Nevertheless, at some time in the future there will almost certainly be a major role played by methodologies that embody features of both disciplines.

Because generalized methodology and statistical inference are both such vast fields, below we will only discuss three specialized topics that can be said to lie within their logical intersection. Also, we will concentrate on hybrid approaches that combine useful features of both disciplines.

The first hybrid approach described below is related to rank-based nonparametric inference. This choice of topics was not made because nonparametric tests are particularly more important than Bayesian and other inferential methodology. It was made because, as their names imply, nonparametric curve estimation and nonparametric testing share common features of generality and decreased dependence upon underlying assumptions. This common viewpoint helps to underscore the potential the newer generalized statistical procedures have to hybridize with classical confirmatory methods.

In this final chapter we will try to dispel an unfortunate myth. As the term model-free which is sometimes used to describe this field implies, this myth asserts that there is an uncrossable chasm that separates conventional elementary-function-based methodology and rank-based procedures. It *is* true that in some places there are clear lines of demarcation. However, consider the following: As discussed in Chapter 2, generalized methods are closely related to pre-1950 Pearson family (PF) and Gram-Charlier-Edgeworth (GCE) approaches, techniques that are clearly parametric while, as shown in the next section, it is possible to modify commonly used classical rank-based statistics, which are often called "nonparametric," by using modern generalized extensions of the parametric PF and GCE approaches. This shows that, at least in this one area, the supposedly uncrossable chasm between rank-based and model-based approaches is largely an illusion. It will be shown that rank-based methods are only one out of a great many, alternatives. The rank-based alternative happens to be the limiting, $m \to \infty$, case of one form of m-coefficient representation. When used for inferential purposes other forms of representation with finite m can serve as intermediates between the two or three parameter elementary function models and the limiting, $m \to \infty$, case.

In Section 10.3 we will briefly turn to questions concerning the interface between Bayesian inference and curve estimation. It will be shown that the posterior density of a location parameter can be expressed in terms of (1) the sample trigonometric moments, (2) the Fourier coefficients of the logarithm of the assumed underlying model, and (3) the Fourier coefficients of the logarithm of the prior density. This relationship illustrates that generalized representation can serve as part of useful, yet largely uninvestigated, Bayesian-based curve estimation and inference hybrids. In addition, it can serve as the representational medium that can expedite the computation of many Bayesian procedures.

10.1 Nonparametric inference and curve estimation

As discussed in previous chapters, penalized likelihood, kernel, as well as series generalized curve estimators, can all be expressed by using the same complete orthogonal system, $\{\exp(2\pi ikx)\}$. It will now be shown that this same form of generalized representation can also be used to implement rank-type "nonparametric" inferential procedures. In this way a wide spectrum of nonparametric inferential and curve estimation hybrids can be constructed. One such hybrid, the generalized two-sample rank test, has power characteristics that will be shown below to match and sometimes even slightly surpass the power of a conventional nonparametric test. That here the improvement in statistical efficiency is small, is attributable to the already highly advanced state of nonparametric inferential methodology, rather than to a fundamental limitation in the potential of curve estimation to eventually contribute to the field of statistical inference.

In contrast to currently available rank-based techniques, the new generalized procedures can easily be extended to the multivariate case. This is because the curve estimators that correspond to the $m \rightarrow \infty$ limiting cases of the statistics described in this section are much harder to generalize to the multivariate case, in terms of computation, than are many other commonly available curve estimation alternatives, and, in particular, the finite m truncated series estimators described in Chapters 3 and 4.

The steps used to derive rank-based inferential methods differ considerably from the steps needed to derive current model-based tests. Model-based inferential procedures are usually formulated in terms of a hypothesis concerning one or more parameters of some model. The problem of sample utilization is a second, albeit important, consideration. On the other hand, rank-based procedures are initially formulated in terms of properties of a ranked sample where, as a first step, it is assumed that data should be used in ranked observation form (Kendall 1962, Hajek and Sidak 1967, and Lehmann 1975).

From a computational point of view ranking is the equivalent of population cumulative estimation by the sample cumulative F^*. In practical terms, for a given iid sample, $\{X_j\}$, $j=1,\ldots,n$, a computer algorithm for sample cumulative step function construction depends on the determination of the rank of each unranked sample component. For example, the rank, u, of the unranked νth member of $\{X_j\}$ satisfies the equation $u=nF^*(X_\nu)$. Therefore, basing a statistical investigation on a rank test is tantamount to the assumption that the estimator F^* within a ranking function nF^* is preferable to any alternative estimator of the population cumulative.

The sample cumulative F^* is, of course, only one of many estimators of the population cumulative F that are available today. The partial integrals of any of the pdf estimators described in Chapter 4 can yield an alternative to F^*. For example, an alternative to F^* can be based on the partial integral of any of the Fourier series pdf estimators of the form

$$\hat{f}(x) = \sum_{k=-m}^{m} b_k \hat{B}_k \exp(2\pi i k x) \tag{10.1.1}$$

where $\{b_k\}$ is any one of the many multiplier sequences discussed in this book. Note that Kronmal and Tarter (1968, Section 3) showed that a partial integral of raw estimator (3.2.5) approaches F^* as $m \to \infty$. It therefore follows that rank-based approaches are limiting cases of the procedures described below, where each individual multiplier $b_k \to 1$, for $k=0,\pm1,\pm2,\ldots$.

Tarter and Lock (1991) showed that alternatives to F^* can be used to generalize rank-based tests by means of the following two step process: (1) Determine the functional of one or more cdfs that corresponds to the limiting case of some particular rank test. (2) Construct alternative estimators by substituting generalized cdf estimator alternatives to F^* for the cdfs within these functionals.

For example, the usual rank correlation statistic r_s computed from a sample of size n (Dixon and Massey, 1983, p. 402) is an estimator of the quantity

$$K_1 - K_2 \mathrm{E}_{f(x,y)}[F_x(X)F_y(Y)], \qquad (10.1.2)$$

where

$$K_1 = 1-2(n+1)(2n+1)/(n-1), \ K_2 = 12n/(n-1),$$

f is the joint density of the variate (X,Y), while F_x and F_y are the marginal cumulatives of the variates X and Y, respectively. For any variates X and Y whose support is on the unit interval and whose means equal one half,

$$\mathrm{E}_{f(x,y)}[F_x(X)F_y(Y)] = \frac{1}{4}\left\{1-\pi^{-2}\sum_{u\neq 0}\sum_{v\neq 0}\left[\frac{(B_{u,0}-1)(B_{0,-v}-1)B_{u,v}}{uv}\right]\right\}. \qquad (10.1.3)$$

Within (10.1.3), $B_{u,v}$ is the u,vth Fourier coefficient of f and the indices of summation range over all nonzero integer values. Suppose that x and y variates are transformed to ensure that variate sample means each equal one half. Then for any $\{\exp(2\pi ikx)\}$-series penalized likelihood or MISE procedure, (10.1.3) can be estimated by substituting an appropriate series coefficient for each $B_{u,v}$.

Any bivariate density estimator that has a convergent Fourier expansion can be used to obtain an estimator of $\mathrm{E}_{f(x,y)}[F_x(X)F_y(Y)]$. For example, for a density estimator based on the kernel K, the product of the u,vth sample trigonometric moment times the u,vth Fourier coefficient of K can be used to estimate any $B_{u,v}$ within (10.1.3). Below, we will collectively refer to test statistics that, like (10.1.3), use this approach as hybrid procedures.

A second example of a hybrid test, which has the rank sum test as a special case, is the following: Let $\overline{X}_1$ and $\overline{X}_2$ designate the sample means of two random samples of size n_1 and n_2 respectively, whose variates have support on the unit interval. Define the statistic T_m as:

$$T_m = n_1 n_2 \left[\overline{X}_1 - \overline{X}_2 + \sum_{k=-m}^{m} \hat{B}_{-k}^{(1)} \hat{B}_k^{(2)} / (2\pi i k) \right] + n_1(n_1+n_2+1)/2, \qquad (10.1.4)$$

where each $\hat{B}_k^{(s)}$, $s=1,2$, is the kth sample Fourier coefficient calculated from the sth sample, s=1,2. Then, Tarter and Lock (1991) showed that as $m \to \infty$, T_m approaches the two sample rank sum test statistic T defined on p. 394 of Dixon and Massey (1983).

There is a simple way to interpret the distributional property that (10.1.4) estimates. Let $f_s, F_s, s=1,2$ represent the pdfs and cdfs of two distributions whose pdfs have support on the unit interval and convergent Fourier expansions with pdf Fourier coefficients $B_k^{(s)}$, $s=1,2$; $k=0,\pm1,\pm2,\ldots$. It follows from integration by parts that

$$\frac{1}{2}[E_{f_1}\{F_2(X)\} - E_{f_2}\{F_1(X)\}] = \frac{1}{2}\left[\int_0^1 F_2(x) f_1(x)dx - \int_0^1 F_1(x) f_2(x)dx\right] \qquad (10.1.5)$$
$$= \mu^{(1)} - \mu^{(2)} + \frac{1}{2\pi i} \sum_{k \neq 0} \frac{B_{-k}^{(1)} B_k^{(2)}}{k},$$

where $\mu^{(1)}$ and $\mu^{(2)}$ are the means of the two populations. Hence, the term within the square brackets of the rank test statistic T_m estimates (10.1.5).

Besides being a crucial part of the relationship between (10.1.5) and T_m, the sum $\sum_{k=-m}^{m} \hat{B}_{-k}^{(1)} \hat{B}_k^{(2)} / (2\pi i k)$ has a second, less direct, connection with the rank test statistics T_m and T. Consider that since a test based on T is not a uniformly most powerful test in the normal case, it is possible for T_m, for some m which is finite, to perform better than the conventional rank-based procedure in the trials described below, which were based on simulated normal data.

Lehmann (1959) Chapter 4, describes the "optimum property of the Wilcoxon test for detecting a shift in the logistic distribution." The logistic density

$$g(x) = e^{-x}(1+e^{-x})^{-2} = \frac{1}{4}\left(\operatorname{sech}\frac{x}{2}\right)^2$$

is fairly similar to the normal in overall shape, but it differs substantially from the normal in terms of its tail characteristics. For example, Tarter (1975b) showed that there does not exist a scale parameter σ of the logistic model such that as $y \to 1$ the inverse cumulative of the logistic, the logit, $\sigma \log[y/(1-y)]$, approaches the standard normal inverse cumulative. Another way of putting this difference between the normal and logistic is that there is no way of scaling the logistic model so that the gap between the logit and the inverse normal remains finite as $y \to 1$. Thus, since the logistic and normal do have very different properties, that the T test is optimal for the logistic does not necessarily make it implausible for T_m to perform better than T for shift alternatives in the normal and other bell-shaped density cases.

10.2 The hybrid Wilcoxon test and the logistic model

T's optimal performance in the logistic case suggests that there may be a connection between (10.1.4) and maximum likelihood estimates of the location parameter of a logistic model. Below this will indeed be shown to be the case.

In Section 8.5 a relationship was discussed that was based on the expression for the sample cdf,

$$F^*(\theta) = \theta - \overline{X} + \frac{1}{2} + (2\pi i)^{-1} \sum_{k \neq 0} \frac{\hat{B}_k}{k} e^{2\pi i k \theta},$$

for $0 \leq \theta \leq 1$. If n is odd, setting θ equal to the median M yields a simple relationship for the difference between the mean and median,

$$\overline{X}-\mathrm{M} = (2\pi i)^{-1}\sum_{k\neq 0}\frac{\hat{B}_k}{\mathrm{k}}e^{2\pi ikM}. \qquad (10.2.1)$$

We will now derive an expession that is almost identical to (10.2.1) where the estimator of the location parameter of a logistic model takes the place of the median. (Note that the median is the ML Laplace model location parameter estimator that was discussed in Chapter 8.)

Consider that

$$F(x) = \int_0^x \sum_{k=-\infty}^{\infty} B_k e^{2\pi ikz}dz = x+(2\pi i)^{-1}\left[\sum_{k=-\infty}^{\infty}(B_k/k)e^{2\pi ikx}-\sum_{k=-\infty}^{\infty}(B_k/k)\right].$$

Suppose a distribution has a pdf f that can be represented as a uniformly convergent Fourier expansion with coefficients B_k, $k=0,\pm 1,\pm 2,\ldots$, a mean μ, and a cdf F. If the difference between the cumulative F and the line $y=x$ can be expressed as a convergent Fourier series, then

$$F(x) = x-\mu+1/2+\frac{1}{2\pi i}\sum_{k\neq 0}\frac{B_k e^{2\pi ikx}}{k}. \qquad (10.2.2)$$

Representing B_k as C_k+iS_k, $k=0,\pm 1,\pm 2,\ldots$, it follows that

$$\frac{1}{2\pi i}\sum_{k\neq 0}\frac{B_k e^{2\pi ikx}}{k} = \frac{1}{\pi}\sum_{k=1}^{\infty}\frac{C_k\sin(2\pi kx)+S_k\cos(2\pi kx)}{k}. \qquad (10.2.3)$$

For the situation where $\{x_j\}$ contains the values assumed by an iid sample of size n of logistic variates $\{X_j\}$, where $G(x-\theta)$ each variate has cdf $G(y)\equiv(1+e^{-(y-1/2)/\sigma})^{-1}$, for $\hat{\theta}$, the location parameter ML estimator of θ, $\hat{\theta}$, satisfies the equation

$$n^{-1}\sum_{j=1}^{n} G(x_j-\hat{\theta}+1/2) = 1/2 \qquad (10.2.4)$$

as discussed by Plackett (1958), Fisk (1961), Tarter and Clark (1965) and Tiku (1968). Assume that σ is sufficiently small to ensure that the truncation resulting from the Fourier series representation of the logistic, which has infinite support, can be ignored. Under this assumption, the substitution of (10.2.2) into (10.2.4) yields the following relationship between $\hat{\theta}$, the sample mean $\overline{X}$, and the sample trigonometric moments $\{\hat{B}_k\}$ computed from $\{x_j\}$:

$$\overline{X}-\hat{\theta} = (2\pi i)^{-1}\sum_{k\neq 0}(-1)^k\frac{B_{-k}^{(1)}\hat{B}_k}{k}e^{2\pi ik\hat{\theta}}, \tag{10.2.5}$$

where $\{B_k^{(1)}\}$ are the Fourier coefficients of the logistic density $f(x)=(\operatorname{sech}(x/2))^2/4$. Here, that for any integer k, $B_{-k}^{(1)}= 2k\pi^2\sigma(-1)^k\operatorname{csch}(2k\pi^2\sigma)$, follows from the relationship for the characteristic function of the logistic with scale parameter and zero location parameter, $E_f(e^{itY})=\pi\sigma t[\operatorname{csch}(\pi\sigma t)]e^{it/2}$ (Johnson and Kotz, 1970). As $\sigma\to 0$, l'Hopital's rule implies that $B_{-k}^{(1)}\to(-1)^k$, which yields the equation

$$\overline{X}-\hat{\theta} = (2\pi i)^{-1}\sum_{k\neq 0}\frac{\hat{B}_k}{k}e^{2\pi ik\hat{\theta}} \tag{10.2.6}$$

that is identical to equation (10.2.1). We have found that an excellent estimator of the logistic scale parameter σ can be obtained by evaluating the logistic $R(x)$ function at, or near, the estimated distribution mode. Hence, when this estimator is substituted into $B_{-k}^{(1)}$, $\hat{\theta}$ can be estimated from (10.2.5). The identity of (10.2.6) and (10.2.1) indicates that the sample median will provide an initial value for the solution of equation (10.2.5). If it is desired that both the location and scale parameters of the logistic be estimated by likelihood procedures, then the equation

$$n^{-1}\sum_{j=1}^{n}\left(\frac{x_i-\hat{\theta}}{\hat{\sigma}}\right)\frac{1-e^{\frac{\pi(x_i-\hat{\theta})}{\hat{\sigma}\sqrt{3}}}}{1+e^{\frac{\pi(x_i-\hat{\theta})}{\hat{\sigma}\sqrt{3}}}}=\frac{\sqrt{3}}{\pi}$$

can be used to either directly estimate the scale parameter ML estimator $\hat{\sigma}$ or check values obtained by $R(x)$ methodology, (Johnson and Kotz, 1970).

Expression (10.2.5) illustrates that the likelihood equation for a logistic location parameter, and the test which is optimum for detecting a shift in the logistic distribution, can be expressed in terms of the nearly identical Fourier coefficients, $B_{-k}^{(1)}\hat{B}_k/(2\pi ik)$ and $B_{-k}^{(1)}\hat{B}_k^{(2)}/(2\pi ik)$. Relationship (10.2.5) can be applied to obtain a new form of logistic regression. By using the same techniques that were described in Chapters 7 and 8, estimated conditional Fourier coefficients can be substituted into both $B_{-k}^{(1)}\hat{B}_k/(2\pi ik)$ and $B_{-k}^{(1)}\hat{B}_k^{(2)}/(2\pi ik)$ in place of the univariate sample trigonometric moments. In this way, (1) a form of semiparametric regression can be implemented that is to the logistic, what the semiparametric LS form of regression curve was for the normal model, or the semiparametric LAV form was for the Laplace model. In addition, as will be described below, the rank test procedure based on T_m can be generalized to obtain a test of the hypothesis that two such curves are coincident.

The following simulation study was performed to study statistic (10.1.4)'s performance. In order to conduct trials in as conservative a fashion as possible, raw estimator series terms were substituted for all $\hat{B}_{-k}^{(1)}$ and $\hat{B}_k^{(2)}$ used to compute T_m. The value of m was chosen by applying the rule, include a kth-term if for the kth coefficients $\hat{B}_{-k}^{(s)}\hat{B}_k^{(s)}>2/(n+1)$, $s=1,2$, $k=0,\pm 1,\pm 2,\ldots$, as described in Tarter and Kronmal (1970). In addition, the simple rule, stop the inclusion process by checking an increasing sequence of indices and stopping once two consecutive terms do not satisfy the above inequality, was applied. Tarter and Kronmal (1976) illustrate the use of this simple procedure. Naturally, as described in Chapters 3 and 4, there are many alternatives to this simple approach, which can, in some circumstances, be used to improve the curve estimator components of hybrid nonparametric tests.

Two studies were conducted. Study (1) consisted of 60 trials designed to compare the numerical values of the two alternatives, T and T_m, when both test statistics were computed from the same sample. Each trial involved two independent iid samples; consisting of $n_1=300$ and $n_2=100$ iid normal variates with variance 1, respectively. Figure 10.1 graphs the test statistics that correspond to the null hypothesis mean pair (μ_1,μ_2) = (0.0,0.0) as small squares. To compare the two statistics T and T_m under the alternative hypothesis, test statistics resulting from trials where the mean pair (μ_1,μ_2) = (0.0,0.2) was chosen are displayed using small circles, and those where the mean pair (μ_1,μ_2) = (0.0,0.4) was chosen are displayed as small triangles. In Figure 10.1 the x-coordinate of each point shown is the value of the series-equivalent of the rank sum based on T_m. The y-coordinate is the conventional rank sum statistic computed

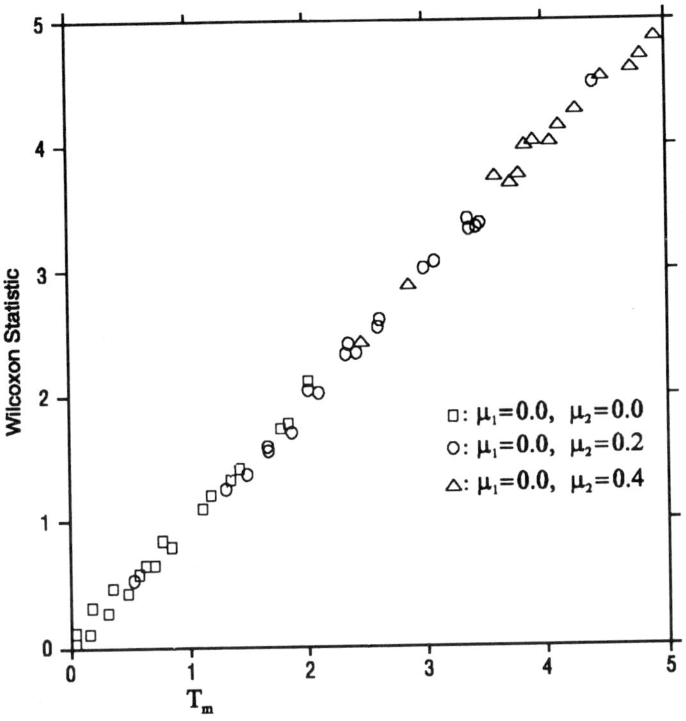

Figure 10.1. *Comparison of series and Wilcoxon test statistics. Each point represents one trial where sample of n_1=300 and n_2=100 points were drawn from normal populations with unit variance and the indicated differences in population means.*

from the same sample. That all points are very close to the diagonal line indicates the near identity between the values of T and T_m in all samples.

Figure 10.2 summarizes the findings of Study 2, which compared the power characteristics of the test statistics T and T_m. Here 300 trials were conducted, where for each trial independent $n_1=300$ and $n_2=100$ samples of iid variates were generated. The near identity of the estimated curves shown in Figure 10.2 suggests that for most applications the power characteristics of tests based on T_m and those based on T are almost identical. For certain alternatives, the hybrid Wilcoxon statistic T_m was found to be slightly more powerful than the conventional T test.

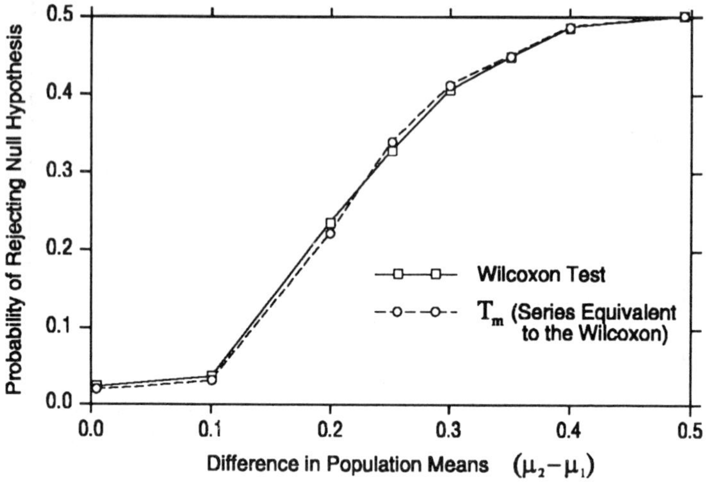

Figure 10.2. *Power curves for the Wilcoxon and its series equivalent. Each point marked on the power curves was derived from 100 trials where the samples were drawn from normal populations with the given difference in population means and unit variances. The sample sizes were* $n_1=300$ *and* $n_2= 100$.

The above finding is in accord with studies described in the last chapter which show that the Kaplan-Meier (1958) estimator, as well as the conventional life-table (Shryock and Siegel, 1973), may be substantially improved by using series estimators based on the metric $J(\hat{f},f,w)$ (Tarter, Freeman, and Polissar, 1990). Note, to be conservative, the statistic T_m was calculated by using one of the simplest of curve estimators, the raw estimator based on a simple two-term stopping rule. This is equivalent to the choice of a Dirichlet KT as the means to combine evaluations of the sample characteristic function. For certain applications it may be useful to choose an alternative to the Dirichlet KT based on the insertion of some nonuniform weight function $w(x)$ in the MISE metric $J(\hat{f},f,w)$, where $w(x)$ increases monotonically. In this way a variant of T_m can be designed that has improved sensitivity to right tail characteristics, for example, has enhanced power to discern long-term survival differences. Ideally, of course, it would be preferable to have the user selection of some curve such as $w(x)$ differentially weight sections of the power curve that assesses the test's performance, rather than use a function like w to input user preferences into $J(\hat{f},f,w)$. Nevertheless, indirect though it may be, the convenience which results from its not requiring that data be ranked as a computational preliminary, as well as the high power performance demonstrated when T_m of (10.1.4) is based on the raw estimator, suggests that this new hybrid test procedure can be of considerable value as an alternative for a conventional nonparametric inferental procedure.

From a general point of view, T_m illustrates only one out of many possible hybrid procedures. Where previously only a single estimator like F^* was used for almost all applications, today nonparametric inferential procedures can be based on a wide variety of alternatives. Besides hybrid versions of conventional univariate tests, there are many possible extensions and multivariate generalizations of hybrid techniques. T_m and its counterparts are easier to generalize beyond the univariate case than are conventional rank-based procedures. For example, as described in Chapter 8, the Fourier coefficients of univariate conditional pdfs can be calculated from the bivariate Fourier coefficients of a two-dimensional pdf. These univariate conditional coefficients can then be used to construct higher-dimensional hybrid inferential procedures.

A particularly useful variant of (10.1.4) is based on the estimators of the Fourier coefficients of separated mixture components described in Chapter 5. For example, suppose it is of interest to discern whether or not the rightmost mixture component shifts from one population to another. Specifically, suppose to assess the effect of exposure to an environmental contaminant such as airborne SO_2, both a control and a diseased population is studied. However, also suppose that, hypothetically, a difference between the control and diseased exposure level populations only occurs when the rightmost mixture component of the control population differs from the rightmost mixture component of the diseased population; while, because within both populations there are individuals whose membership in the diseased or the control population has nothing at all to do with SO_2 exposure, the two populations have identical leftmost components.

When two independently distributed random samples are available from the control and from the diseased population respectively, then methods described in Chapter 5 can be applied to construct two independently distributed sequences of estimators of the Fourier coefficients corresponding to the rightmost subcomponents of two populations. The and $\hat{B}_{-k}^{(1)}$ and $\hat{B}_k^{(2)}$, $k=0,\pm 1,\pm 2,\ldots$ components of (10.1.4) can then be approximated by these sequences. Here, appropriate values of n_1 and n_2 can be approximated by multiplying the estimated mixing parameter of each rightmost component times the number of control and disease measurements respectively. Finally, the $\overline{X}_1$ as well as the $\overline{X}_2$ components of (10.1.4) can be approximated by using the expression for the mean in terms of estimated Fourier coefficients that was given in Chapter 8. Consequently, a shift test based on the esimation of (10.1.5) can be constructed for the null hypothesis that there is no difference between the rightmost components of the two subpopulations, for a nonzero rightmost component shift alternative. The advantages of this test stem from its capacity to reduce the confounding that may be tractable to the existence of two classes of mixture components, the first not differentiated by control versus disease population membership, and the second, which do shift. Unless this confounding is controlled, the capacity to discern differences which, as is likely in practice, only apply to rightmost or other specific distribution subpopulations, will be greatly reduced. This confounding is likely to be encountered whenever the presence of an individual in a diseased group can be attributed to factors other than exposure to the environmental contaminant whose levels are under study.

The estimates of (10.15) depend on the two sample sizes n_1 and n_2 as well as the sample trigonometric moments. As indicated above, in mixture decomposition applications, values of n_1 and n_2 can be determined by estimating the mixing parameters of components and multiplying these estimates by the total sample sizes that were used to decompose the mixture. For the case of applications that are based on the conditional Fourier coefficients the following more general approach can be used: Consider the linear combination, or weighted sum, (10.2.7) of the multivariate sample trigonometric moments

$$\sum_{k_1=-m_1}^{m_1}\sum_{k_2=-m_2}^{m_2}\cdots\sum_{k_q=-m_q}^{m_q}\alpha_{(k_1,\ldots,k_q)}\hat{B}_{(k_1,\ldots,k_q)}, \tag{10.2.7}$$

which, for notational convenience, will be expressed as $\sum_{u=-m}^{m}\alpha_u\hat{B}_u$.

From relationships described in Tarter (1970) the variance of (10.2.7) can be shown to equal

$$(1/n)\sum_{u=-m}^{m}\sum_{v=-m}^{m}\alpha_u\alpha_v(B_{u+v}-B_uB_{-v}) \tag{10.2.8}$$

where the index u-v represents the q-tuple whose sth element is the difference between the sth element of u and the sth element of v. The expression

$$\sum_{u=-m}^{m}\sum_{v=-m}^{m}\alpha_u\alpha_v\left(\frac{n+2}{n(n+1)}\hat{B}_{u-v}-\hat{B}_u\hat{B}_{-v}/(n+1)\right) \qquad (10.2.9)$$

can be shown to be an unbiased estimator of (10.2.8).

Through use of (10.2.9) trial values of n_1 and n_2 can be obtained for the generalization of T_m where the estimated Fourier coefficients of two conditional densities are used to replace univariate trigonometric moments. The basic idea that underlies this approach is to estimate the size of sample from which the sample trigonometric moments would have a variance estimated by (10.2.9).

Specifically, let n represent the size of the entire bivariate sample. By substituting $\alpha_u \doteq e^{2\pi iux_0}/f(x_0)$, $u=0,\pm1,\pm2,\ldots$, into (10.2.9), the variance of the estimator of each $B_v^{(x_0)}$, $v=0,\pm1,\pm2,\ldots$ within the conditional

$$\sum_{v=-\infty}^{\infty} B_v^{(x_0)}e^{2\pi ivy},$$

where

$$B_v^{(x_0)} = \left[\sum_{s=-\infty}^{\infty} B_{s,v}e^{2\pi isx_0}\right]\Big/ f(x_0)$$

can be approximated. Below, this quantity will be represented as $\hat{\sigma}_v^2$. Let $\hat{n}_v$ represent the estimate of the sample size of a univariate sample which would have variance $\hat{\sigma}_v^2$. Since the variance of a univariate Fourier coefficient whose modulus equals β is $(1/n)[1-\beta]$, a reasonable choice of $\hat{n}_v$ is $(1-|B_v^{(x_0)}|^2)/\hat{\sigma}_v^2$. It is best to use some weighted average of $\hat{n}_v$ where weights decrease rapidly.

10.3 Bayesian inference and curve estimation

Generalized methodology can be applied to confirmatory studies based on Bayes' Theorem. In 1958 Whittle discussed what he called, the "smoothing of probability density functions," from a Bayesian point of view. However, the Whittle, as well as related studies by Fellner (1974) and Brunk (1978), concentrate on exploratory rather than confirmatory applications of Bayesian methods.

In this section Lindley's (1965) notation will be used to represent: (1) likelihood, the "density in the frequency sense," $\pi(x|\theta,\mathrm{H})$, where H denotes our state of knowledge before the sample was taken, and θ represents "a parameter of the family of distributions defined by the densities $\{f(x|\theta)\}$," (2) the prior distribution $\pi(\theta|\mathrm{H})$, and (3) the posterior density of θ, $\pi(x|\theta,\mathrm{H})$.

Suppose a Bayesian approach is based on the underlying model g and it is assumed that over the unit interval the Fourier coefficients of $\log[g(z)]$ are well-defined. Also, it is assumed that each of n iid variates has a density $f(x|\theta)\equiv g(x-\theta)$, where the uniformly convergent Fourier expansion of f has coefficients $\{B_k\}$, and the natural log of $\pi(x|\theta,\mathrm{H})$ can be expressed in terms of the sequence of Fourier coefficients $\{\beta_k\}$ of the natural log of model $g(z)$. In Section 8.2, it was shown that

$$\log\pi(x|\theta,\mathrm{H}) = \mathrm{n}\sum_{k=-\infty}^{\infty} \hat{B}_k\beta_k \exp(2\pi ik\theta).$$

Consequently, if the log of the prior distribution of θ has the Fourier expansion

$$\sum_{k=-\infty}^{\infty} c_k\beta_k \exp(2\pi ik\theta),$$

then the posterior density of θ is

$$\pi(x|\theta,\mathrm{H}) = \exp\left\{n\sum_{k=-\infty}^{\infty} [\hat{B}_k+(c_k/n)]\beta_k \exp(2\pi ik\theta)+C\right\} \quad (10.3.1)$$

where the constant C is not functionally related to θ.

Formula (10.3.1) would be identical to the expression for the likelihood function of a location parameter presented in Section 8.2 were it not that the term $[\hat{B}_k+(c_k/n)]$ replaces each of the sample trigonometric moments $\hat{B}_k$, which appear in the coefficient of the Fourier series location parameter MLE equation. However, for a prior that is uniform on the unit interval, for all $k\neq 0$, c_k equals 0. Thus, for a uniform prior, $\pi(x|\theta,\text{H})$ is identical to Fourier series location parameter MLE equation discussed in Chapter 8.

The uniqueness of Fourier series expansions can help answer the following question: Can two different pairings of prior and model result in the same location parameter posterior distribution? The reason this question is important is that unlike most of the procedures described previously in this book, Bayesian inference is based on assumptions concerning two curves, the model and the prior, rather than a single curve, such as the $w(x)$ component of MISE $J(\hat{f},f,w)$. If two curves, for example, $f(x)$ and $\pi(\theta,\text{H})$, are zero outside some interval of finite length, it is easy to construct a single curve from which both $f(x)$ and $\pi(\theta,\text{H})$ can be determined. However, in Bayesian statistical applications the assumption of a model and the assumption of a prior are conventionally interpreted as involving two distinct curves, and the use of a single assumed curve, one which combines the role of model and prior, has never been advocated.

We will now show that at least for the case of location parameter estimation, $f(x)$ and $\pi(\theta,\text{H})$ distinctiveness corresponds to the uniqueness of a model, prior, pairing. Specifically, there cannot exist two distinct pairings of model and prior that will always yield the same posterior density $\pi(x|\theta,\text{H})$ for any sample.

Model-prior uniqueness can be deduced if, when calculated from different samples, the sample trigonometric moments obtained from an iid sample of data with the above datum pdf $f(x)$ can take on at least two distinct values, say for the kth coefficient, $\hat{B}_k\neq\hat{B}'_k$. In practice, model-prior uniqueness implies that even though $f(x)$ and $\pi(\theta,\text{H})$ might be combined to express the assumptions used in Bayesian analysis in terms of the properties of a single curve, this curve would still have to be separated into its $f(x)$ and $\pi(\theta,\text{H})$ components before the posterior density $\pi(x|\theta,\text{H})$ was determined. In this sense, it is intrinsic to the Bayesian approach that assumptions concerning two curves, and not the assumption of a single curve like $w(x)$, be made.

The proof of the location uniqueness relationship is as follows: Suppose $\{g^{(s)}(z),\pi^{(s)}(\theta)\}$, $s=1,2$ designate two model and prior pairs. From the Fourier series uniqueness theorem, that $g^{(1)}(z)$ is not identically equal to $g^{(2)}(z)$, implies that the kth Fourier coefficient $\beta_k^{(1)}$ of $\log[g^{(1)}(z)]$ cannot equal the kth Fourier coefficient of $\beta_k^{(2)}$ $\log[g^{(2)}(z)]$ for at least one value of the index, specifically, k. Let $c_k^{(1)}\beta_k^{(1)}$ represent the kth Fourier coefficient of log-prior $\ln[\pi^{(1)}(\theta)]$ and $\beta_k^{(2)}$ represent the kth Fourier coefficient of log-prior $\ln[\pi^{(2)}(\theta)]$. In addition, suppose it were true that when computed from two different samples, $\{g^{(s)}(z),\pi^{(s)}(\theta)\}, j=1,2$, yield identical posterior densities of the form (10.3.1). From the uniqueness theorem of Fourier series it follows that

$$[\hat{B}_k+c_k^{(1)}/n]\beta_k^{(1)} = [\hat{B}_k+c_k^{(2)}/n]\beta_k^{(2)}.$$

and

$$[\hat{B}_k'+c_k^{(1)}/n]\beta_k^{(1)} = [\hat{B}_k'+c_k^{(2)}/n]\beta_k^{(2)}.$$

These simultaneous equations imply that

$$[\hat{B}_k-\hat{B}_k']\beta_k^{(1)} = [\hat{B}_k-\hat{B}_k']\beta_k^{(2)},$$

which, since $\hat{B}_k\neq\hat{B}_k'$, also implies that $\beta_k^{(1)}=\beta_k^{(2)}$. The two simultaneous equations also imply $c_k^{(1)}=c_k^{(2)}$, which proves model-prior uniqueness.

The above uniqueness proof concerns the entire posterior density $\pi(x|\theta,\mathrm{H})$ of an estimator. The partial derivatives of $\ln\pi(x|\theta,\mathrm{H})$ with respect to elements of the vector θ can be taken to deal with the problem of whether or not two distinct models can yield the same likelihood estimator. For example, consider the one-parameter case of the uniform prior where $c_k=0$ for $k>0$, and where the log likelihood

$$\log\pi(x|\theta,\mathrm{H}) = n\sum_{k=-\infty}^{\infty}[\hat{B}_k+(c_k/n)]\beta_k\exp(2\pi ik\theta)+C \qquad (10.3.2)$$

has a uniformly convergent derivative with respect to θ. The constant C is not functionally related to the parameter θ. Therefore, for any sample of size n of iid variates with datum density $f(x|\theta)\equiv g(x-\theta)$, the sample trigonometric moments $\{\hat{B}_k\}$, and ML estimator $\hat{\theta}$, satisfy the equation

$$\sum_{k=-\infty}^{\infty} \hat{B}_k k \beta_k \exp(2\pi i k \hat{\theta}) = 0 \tag{10.3.3}$$

whenever the left side of this equation converges.

The solution of simultaneous equations that contain an infinite number of unknowns involves difficult theoretical and computational considerations. Nevertheless, at least conceptually, (10.3.3) suggests the following interesting problem: For the uniform prior case, will a specified statistic, for example, the sample mean $\overline{X}$, be found to be a location parameter likelihood estimator $\hat{\theta}$ under the assumption of any, only one, or c distinct underlying models g_s, $s=1,\ldots,c$?

The above question can be interpreted as being the opposite of the usual likelihood approach which begins with a choice of standard model and ends with the computation of a ML estimator. Many statistics have been suggested as location parameter estimators based on other than likelihood or Bayesian criteria. For example, there are many robust location estimators. However, it could easily be the case that just as the sample median, which happens to be both a simple, and moderately robust, estimator, is the ML estimator of a Laplace model location parameter, other robust statistics also happen to be ML estimators of the location parameter of a particular model (or of several particular models): But *which* particular model (or models)?

An approximate solution to this problem can be based on the assumption that $\beta_k=0$, for $k=\pm(m+1),\pm(m+2),\ldots$ Under this assumption, $2m$ samples will yield a system of $2m$ linear homogeneous equations, each of form (10.3.3). Samples need not necessarily be random although it should be taken into account that some trial samples may yield redundant equations. Under the constraint that the solution of this system corresponds to a sequence of Fourier coefficients of the logarithm of some density g, if this system of linear equations yields a unique solution for $\{\beta_k\}$, then the g thus determined is the unique density that corresponds to the particular ML estimator of location parameter θ. Hence, the likelihood equations are used in the reverse of their usual application—not to relate a given choice of g to a statistic like $\overline{X}$—but instead to see whether a specific statistic like $\overline{X}$ can be the location likelihood estimator based on the assumption of more than a single model g.

Note that the association of the sample mean with both the normal and Poisson densities does not imply that there are two systems of equations based on (10.3.3) that have the sample mean as a common solution. This is because the sample mean is not the ML estimator of the *location* parameter of the Poisson density. Consequently, a system of equations different from linear equations based on (10.3.3) relates the sample mean to the Poisson model and the sample trigonometric moments.

10.4 The Watson A-test and its generalizations

Besides hybrid procedures that share features of curve estimation methods with either rank-based, or Bayesian, techniques, there are other hypothesis-testing applications of generalized methodology. One early and valuable procedure concerns a test for the uniformity of a circular distribution (Watson, 1967; Parzen, 1977). Watson's A-test has a close association with the estimated MISE $J(\hat{f},f)$ of a density estimator $\hat{f}$ based on the specialized form of kernel K described in this section. Here, $\hat{f}$ estimates a density f hypothesized to have zero Fourier coefficients for indices k that are nonzero. Watson observed that when Fourier coefficients $\hat{B}_k, k=0,\pm1,\pm2,\ldots$, are calculated from iid data, "as n approached infinity [these] Fourier coefficients become normally distributed and independent." Because the null hypothesis of uniformity determines the values of Fourier coefficient sequence $\{B_k\}$, the values of the variances of $\{\hat{B}_k\}$ are constant and known, while the covariance of any two distinct members of $\{\hat{B}_k\}$ is zero. By making use of these properties of the sample trigonometric moments, he was able to apply an estimator, which he denoted A, of the MISE $J(\hat{f},f)$ to test for uniformity.

Watson's (1967) derivation of the A-test begins with the definition of a statistic that he represented as N(x). N(x) equals the number of sample points on the perimeter of a semicircular portion of a circle of unit circumference, where the semicircle begins at x and proceeds in a clockwise direction. It is convenient to describe the reasoning that underlies Watson's test in terms of a specialized form of univariate kernel estimator. This derivation avoids the requirement that points distributed on a circle, rather than points distributed on a line, be considered.

Suppose n iid variates $\{X_j\}$, $j=1,\ldots n$, are on the half-open interval $[-\pi,\pi)$. Define the kernel K as

$$K(z) = \sum_{u=-\infty}^{\infty} I_{[u\pi-\pi/2,u\pi+\pi/2]}(z). \tag{10.4.1}$$

The kernel K plays the same role over the interval $[-\pi,\pi)$ that a uniform kernel with height equal one plays on a circle of circumference 2π. But, unlike conventional kernels defined over a line, if a portion of the kernel centered about a given data point protrudes beyond the endpoints of the interval $[-\pi,\pi)$, then this portion reappears at the other end of the interval.

Kernel K's kth Fourier coefficient b_k equals the integral from $-\pi/2$ to $\pi/2$ of the function $(1/2\pi)\exp(-ikx)$. Hence, b_k equals zero for even k, and equals $[-1/(\pi k)](-1)^{(k+1)/2}$ for odd k. Define the statistic

$$N(x) = n/2+\sum_{k=-\infty}^{\infty} b_k\hat{B}_k\exp(2\pi ikx),$$

where $\{\hat{B}_k\}$ are the sample trigonometric moments of $\{X_j\}$ $j=1,\ldots,n$. To determine the distribution of A under the null hypothesis of uniformity, Watson observed that, when computed from a random sample, the sample Fourier coefficients are sample means of independent variates. Hence they approach normal variates and the sample trigonometric moment covariance formula used to obtain (10.4.8) can be applied to $\{\hat{B}_k\}$ where, under the null hypothesis of uniformity, all but the zeroth-order population Fourier coefficients of the sequence $\{B_k\}$ equal zero.

Watson's test criterion A is the estimated MISE $J(\hat{f},f)$ divided by the sample size n, where f is the uniform density and $\hat{f}$ is the estimator of f determined by the multipliers $\{b_k\}$, where the sequence $\{b_k\}$ consists of equally spaced evaluations of the KT of the kernel K used to define N(x). It can be shown using Parseval's theorem that the statistic A equals the sum over k of the squared moduli of quantities formed by multiplying the kth respective sample trigonometric moment by the kth-term of the sequence $\{b_k\}$. Watson (1967) used the asymptotic normality of the sample trigonometric moments to determine the limiting distribution of A under the null-hypothesis of uniformity.

The Watson-A procedure closely resembles several other applications of generalized methodology to statistical inference. One such example is the Hodges-Lehmann technique based on a linear score function (Lehmann, 1963), which is discussed in a paper by Schweder (1975). The Hodges-Lehmann approach depends on the calculation of the integral of the squared density that is identical to the functional $J(f,0)$. This quantity, which appears in expressions (3.4) and (3.11) of Lehmann (1963), plays a pivotal role in what Lehmann calls "asymptotically nonparametric inference." If the support of f is on the unit interval, (0,1), then $J(f,1)$, the quantity estimated by Watson's A, equals $J(f,0)-1$, in other words, the Hodges-Lehmann functional minus one. Thus it is natural to apply the same technique that was introduced by Watson to asymptotically nonparametric inference since, to all intents and purposes, estimators of both the Hodges-Lehman functional, and Watson's A, estimate the same distributional property.

A second generalization of A is an extensive and useful series of applications of Fourier series methods to distributional hypothesis testing derived by Langholz (1984, 1987, 1991). In this work it is shown that the general approach used to derive Watson's A can be applied to deal with testing procedures involving a wide variety of underlying distributions and nuisance parameters.

The Langholz method is based on Neyman's (1959) "smooth test" for goodness-of-fit as generalized by Thomas and Pierce (1979). Other procedures for using generalized methodology for the purposes of goodness-of-fit assessment and testing for symmetry about the origin are given by Paulson and Thornton (1975) and Feuerverger and Mureika (1977). Closely related and useful testing procedures based on the Hermite complete system are described by Schwartz (1969a,b).

Significance level determination for all the above tests can be based on Watson's (1967) application of $\{\hat{B}_k\}$ member asymptotic normality. Similarly, the asymptotic normality of $\{\hat{B}_k\}$ members was used by Tarter and Marshall (1978) for a Fourier series variant of a bootstrap procedure. The Fourier bootstrap method uses series representation of statistics that are computed from a sample of size n of either normal or, alternatively, nonnormal variates. Even when the latter forms of variates are studied, trials based on q random normal variates, where q is substantially less than n, can be conducted. For example, many statistics, including all location parameter likelihood estimators discussed in the logistic case in this chapter, and the Pearson family and Laplace case in Chapter

8, can be expressed by a simple relationship involving the sample trigonometric moments $\{\hat{B}_k\}$.

By using Watson's approach to the determination of the asymptotic distribution of A, it can be shown that the vector composed of a finite number of members of the sequence $\{\hat{B}_k\}$ will approach a multivariate vector-valued normal variable with both a mean vector and a variance-covariance matrix expressible in terms of $\{B_k\}$. Thus, for a moderate to large iid sample of variates with datum density f, it is only necessary to know the Fourier expansion of f, and additionally have access to a *normal* random number generator, to generate a sequence of estimators of $\{B_k\}$ with distributional properties almost identical to those of $\{\hat{B}_k\}$ computed from an actual sample of n iid variates with density f. These estimators $\{\hat{B}_k\}$ can in turn be used to compute the desired statistic $\hat{\theta}$.

For example, consider the case where simulated trials of the statistic $s^2 = n^{-1}\sum_{j=1}^{n}(X_j - \overline{X})^2$ or the sample absolute deviation $\eta = n^{-1}\sum_{j=1}^{n}|X_j - \mathrm{M}|$ are conducted, where $\overline{X}$ is the sample mean and M is the sample median. As described in Tarter and Marshall (1978) if all data points are transformed to within the interval $[(3\alpha_1+\alpha_2)/4,(3\alpha_2+\alpha_1)/4]$ (these limits are attributable to Brunk (1978)), then for $C=\alpha_2-\alpha_1$, it follows that

$$s^2/C^2 = \frac{1}{12} + C\sum_{k\neq 0}\left\{\frac{\hat{B}_k(-1)^k\exp[2\pi ik((\overline{X}-\alpha_1)/C)]}{2\pi^2k^2}\right\} \quad (10.4.2)$$

and

$$\eta/C = \frac{1}{4} + C\sum_{k\neq 0}\left\{\frac{\hat{B}_k[(-1)^k-1]\exp[2\pi ik((\mathrm{M}-\alpha_1)/C)]}{2\pi^2k^2}\right\}. \quad (10.4.3)$$

In addition to (10.4.2) and (10.4.3), as discussed in Chapter 8, both the sample mean and the sample median can be computed using formulas that, like (10.4.2) and (10.4.3), are linear in terms of the sample trigonometric moments $\{\hat{B}_k\}$. Therefore, by using *normal* random deviates to generate variates that, because of the Central Limit Theorem, have the same limiting distribution as $\{\hat{B}_k\}$, it is possible to simulate iid samples of the four statistics, $\overline{X}$, M, s^2 and η, associated with any sample of variates whose density, normal or nonnormal, has specified Fourier coefficients $\{B_k\}$.

This simulation method can be applied to test features of the curve estimation process itself. It is shown in Tarter and Marshall (1978) that by using only 14 normal random deviates per pseudosimulated sample of size from 25 to 500, the mean of ISE densities can be approximated to four digit accuracy. It was also shown that the modes of multimodal ISE densities can be accurately estimated.

For many applications the series-based-bootstrap method can be used to corroborate exploratory findings in the following two ways:

(1) Suppose a researcher is interested in a specific curve, or in one of its properties, and for a given sample, $\{\hat{B}_k\}$ are calculated. These sample trigonometric moments can be used to estimate the expectations, variances and covariances of a new pseudo-$\{\hat{B}_k\}$ sequence, which in turn can be used to obtain a pseudo-curve or pseudo-curve property. If the feature of the curve or curve property that the researcher is studying reappears in most pseudo-trials, then the researcher will have tentative assurance that the feature is not an anomaly attributable to sampling variation.

(2) The curve or property associated with the given $\{\hat{B}_k\}$ sequence can be simulated using a pseudo-$\{\hat{B}_k\}$ sequence generated by using the above methodology, but with user specification of the distributional structure of the underlying sample elements by specification of the sequence $\{B_k\}$ and user specification of sample size n.

Note, unlike conventional simulation methodology, both procedures (1) and (2) are exclusively based on a *normal* generator. Like other bootstrap procedures, pseudo-samples are utilized. However, these pseudosamples do not necessarily consist of variates whose density coincides with the datum density. Instead, by taking advantage of the asymptotic normality of the sample trigonometric moments the pseudosampling process can utilize simulated *normal* variates to generate statistics calculated from samples of iid variates that have *any* specified distribution. In addition, while data domain simulation procedures place time demands that increase linearly with sample size, frequency or Fourier domain bootstrap accuracy improves as sample size increases, even though the number of normal variates required increases slowly with increasing n. Hence, the statistical advantage of using a large sample size n can be purchased at an acceptable price of computer time.

In light of the many opportunities, like separating mixture components while avoiding overlap, afforded by data *analysis* in the frequency domain, it is not surprising that the opposite of data analysis, data *simulation*, can often be accomplished expeditiously in the frequency domain.

References

Abramowitz, M. and Stegun, I. A. (1964). *Handbook of Mathematical Functions with Formulas, Graphs and Mathematical Tables*. National Bureau of Standards: Applied Mathematics Series 55.

Anderson, G.D. (1969). *A Comparison of Methods of Estimating a Probability Density Function*. Unpublished Doctoral Dissertation, University of Washington.

Anderson, G.L. and de Figueiredo, R.J.P. (1980). An Adaptive Orthogonal-series Estimator for Probability Density Functions. *The Annals of Statistics* **8**; 347–376.

Barlow, R.E. and Van Zwet, W.R. (1970). Asymptotic Properties of Isotonic Estimators for the Generalized Failure Rate Function, Part II. In *Nonparametric Techniques in Statistical Inference*, Ed. Puri, M.L. Cambridge: Cambridge University Press, 159–173.

Barnett, V. and Lewis, T. (1984). *Outliers in Statistical Data,* New York: John Wiley and Sons.

Bartholomew, D. J. (1957). A Problem in Life Testing. *Journal of the American Statistical Association* **52**; 350–55.

Bartlett, M. S. (1963). Statistical estimation of density functions. *Sankhya, Series A* **25**; 245–254.

Bechtoldt, H .P. (1958). Selection. In *Handbook of Experimental Psychology*. Ed. Stevens, S.S. New York: Wiley. 1237–1266.

Bernstein, S. (1912). Demonstration du Theoreme de Weierstrass, Fondee sur le Calcul des Probabilites. *Commun. Soc. Math. Kharkow (2)* **13**; 1–2.

Bhattacharya. C.G. (1967). A Simple Method of Resolution of a Distribution into Gaussian Components. *Biometrics* **23**; 115–135.

Bjerve, S. and Doksum, K. (1992). Correlation Curves: Measures of Association as Functions of Covariate Values. *Annals of Statistics* (to be published).

Blaydon, C. C. (1967). Approximation of Distribution and Density Functions. *Proc. IEEE* **55**; 231–232.

Boas, R.P., (1949). Representation of Probability Distributions by Charlier Series. *Annals of Math. Stat.* **20**; 376–392.

Bowman, A.W. (1984). An Alternative Method of Cross-validation for the Smoothing of Density Estimates. *Biometrika* **71**; 353–360.

Bowman, A.W. (1985). A Comparative Study of Some Kernel-based Nonparametric Density Estimators. *Journal of Statistical Computation and Simulation* **21**; 313–327.

Breiman, L., Meisel, W., and Purcell, E. (1977). Variable Kernel Estimates of Multivariate Densities. *Technometrics* **19**; 135–144.

Brinton, W. C. (1917). *Graphical Methods for Presenting Facts*. New York. The Engineering Magazine Company.

Brunk, H. D. (1965). Conditional expectation given a σ-lattice and applications. *Ann. Math. Statist.* **36**; 1339–1350.

Brunk, H.D. (1977). Personal communication regarding: Tarter, M.E. and Marshall, J. (1978). New Procedure Designed for Simulation Studies. *Communications in Statistics-Series B* **7**; 283–93.

Brunk, H.D. (1978). Univariate Density Estimation by Orthogonal Series. *Biometrika* **65**; 521–528.

Brunk, H.D. and Pierce, D.A. (1974). Estimation of discrete densities for computer-aided differential diagnosis of disease. *Biometrika* **61**; 493–501.

Brunk, H.D., Thomas, D.R., Elashoff, R. and Zippin, C., (1975). Computer-aided Prognosis. in *Perspectives in Biometrics*, ed. R.M. Elashoff, New York. Academic Press, 63–80.

Buchanan-Wollaston, H.G. and Hodgeson, W.C. (1929). A New Method for Treating Frequency Curves in Fishery Statistics, with some Results. *Journal of Conservation* **4**; 207–225.

Burton, R. F. (1886). *The Arabian Nights' Entertainments* (Editor, B. A. Cerf. Modern Library 1959), New York, Modern Library.

Butler, W.J. (1980). Classification Procedures for Polychotomous Random Variables Using Orthogonal Functions. Unpublished Ph.D. disertation. University of Washington. Dept. of Biostatistics.

Butler, W.J., Kronmal, R.A. (1985). Discrimination with Polychotomous Predictor Variables Using Orthogonal Functions. *Journal of the American Statistical Association* **80**; 443–448.

Carleson, L. (1966). On Convergence and Growth of Partial Sums of Fourier Series. *Acta Math.* **116**; 135–137.

Cassie, R.M. (1954). Some Uses of Probability Paper in the Analysis of Size Frequency Distributions. *Australian Journal of Marine and Freshwater Research* **5**; 513–522.

Cencov, N. N. (1962). Evaluations of a unknown distribution density from observations. *Soviet Math.* **3**; 1559–62.

Charlier, C.V.L. and Wicksell, S.D. (1924). On the Dissection of Frequencey Functions. *Ark. Math. Astr. Fys.* **18**; 1–64.

Cicchinelli, A.L. (1962). *The Composite of Two Gaussian Distributions as a*

Model for Blood Pressure Distributions in Man. Unpublished Ph.D. dissertation, University of Michigan Department of Biostatistics.

Cohen, A.C. Jr. (1988). Censored, Truncated, and Grouped Estimators. In *Lognormal Distributions*. Ed. Crow, E.L. and Shimizo. K. New York: Dekker, Chapter **5**; 139–172.

Cohen, A.C. Jr. (1951). Estimating Parameters of Logarithmic-Normal Distributions by Maximum Likelihood, *Journal of the American Statistical Association* **46**; 206–212.

Craig, C.C. (1936). Sheppard's Corrections for a Discrete Variable. *Annals of Mathematical Statistics* **7**; 55.

Crain, B.R. (1973). A Note on Density Estimation Using Orthogonal Expansions. *Journal of the American Statistical Association* **68**; 964–965.

Crow, E.L. and Shimizo. K. (1988). *Lognormal distributions*. New York: Dekker.

Daniel, C. (1959). Use of Half-normal Plots in Interpreting Factorial Low-level Experiments. *Technometrics* **1**; 311–341.

Dewey (1938). *Logic: The Theory of Inquiry*. New York: Henry Holt.

Devroye, L. (1987). *A Course in Density Estimation*. Boston: Birkhauser.

Devroye, L. and Gyorfi, L. (1985). *Nonparametric Density Estimation: The* λ_1 *View*. New York: John Wiley and Sons.

Diggle, P.J. and Hall, P. (1986). The Selection of Terms in an Orthogonal Series Density Estimator. *Journal of the American Statistical Association* **81**; 230–233.

Dixon, W.J. and Massey, F.J. (1983). *Introduction to Statistical Analysis*, New York: McGraw-Hill.

Doetsch, G. (1928). Die Elimination des Dopplereffekts auf Spektroskopische Feinstrukturen und Exakte Bestimmung der Komponenten. *Zs. Phys.* **49**; 705–730.

Doetsch, G. (1936). Zerlegung einer function in Gauss'sche fehlerkurven. *Mathematische Zeitschrift* **41**; 283–318.

Doksum, K.A., Blyth, S., Bradlow, E., and Meng, X. (1991). Correlation curves as local measures of variance explained by regression. Technical Report, Department of Statistics, Harvard University.

Doksum, K.A., Blyth, S., Bradlow, E., Meng, X., and Zhao, H. (1993) Correlation curves as local measures of variance explained by regression. *Journal of the American Statistical Association* (to be published).

Doksum, K.A. and Samarov, A. (1991). *Measures of association in nonlinear regression models*. Technical Report, Department of Statistics, Harvard University.

Duin, R.P.W. (1976). On the Choice of Smoothing Parameters for Parzen Estimators of Probability Density Functions. *IEEE Transactions on Computers* **C–25**; 1175–1179.

Dym, H. and McKean, H.P. (1972). *Fourier Series and Integrals*. New York: Academic Press.

Edwards, R.E. (1967). *Fourier Series: A Modern Introduction*. New York: Holt Rinehart.

Eisenberger, I. (1964). Genesis of Bimodal Distributions. *Technometrics* **6**; 357–363.

Elderton, W.P. (1953). *Frequency Curves and Correlation*. Washington, D.C.: Harren Press.

Feigl, P. and Zelen, M. (1965). Estimation of Exponential Survival Probabilites with Concomitant Information. *Biometrics* **21**; 826–838.

Fellner, W.H. (1974). Heuristic Estimation of Probability Densities. *Biometrika* **61**; 485–492.

Fellner, W.H. and Tarter, M.E. (1971). Some New Results Concerning Density Estimates Based upon Fourier Series. In *Proceedings of Computer Science and Statistics: 5th Annual Symposium on the Interface*. North Hollywood: Western Periodicals.

Feuerverger, A. and Mureika, R.A. (1977). The Empirical Characteristic Function and Its Applications. *The Annals of Statistics* **5**; 88–97.

Fisher, R.A. (1922). On the Mathematical Foundations of Theoretical Statistics. *Philosophical Transactions of the Royal Society of London, Series A* **222**; 309–368.

Fisk, P.R. (1961). Estimation of Location and Scale Parameters in a Truncated Grouped Sech Square Distribution. *Journal of the American Statistical Association* **56**; 692–702.

Fisz M. (1963). *Probability Theory and Mathematical Statistics*. New York: John Wiley and Sons.

Fix, E. and Hodges, J. L. (1951). Non-parametric discrimination: consistency properties, *U. S. School of Aviation Medicine, Proj.* 21–49–004, Rep. 4. Randolph Field, Texas.

Frankowski, R. (1967). *The Estimation of Parameters in a Mixture of Gaussian Distributions with Applications to the Distribution of Serum Uric Acid in Man*. Unpublished Doctoral dissertation, University of Michigan Department of Biostatistics.

Freeman, M.F. and Tukey, J.W. (1949). Transformations Related to the Angular and the Square Root. *Memorandum Report 25*. Statistical Research Group: Princeton Univeristy.

Freeman, M.F. and Tukey, J.W. (1950). Transformations Related to the Angular and the Square Root. *Annals of Mathematical Statistics* **21**; 607–611.

Fukunaga, K. and Ito, T. (1963). A Design Theory of Recognition Functions in Self-organizing Systems. *Trans. Inst. Electron. Commun. Eng. Jap.* **46**; 1672–1679.

Gallant, A. R. and Fuller, W. A. (1973). Fitting segmented polynomial regression models whose join points have to be estimated. *Journal of the American Statistical Association* **68**; 144–147.

Gilchrist, W.G. and Jewell, N.P. (1982). Some Multiple Q-Q Plotting Procedures. In *Modern Data Analysis*, Eds. Launer and Siegel. New York: Academic Press.

Good, I.J. and Gaskins, R.A. (1972). Global nonparametric estimation of probability densities. *Virginia Journal of Science* **23**; 171–193.

Good, I.J. and Gaskins, R.A. (1980). Density Estimation and Bump-hunting by the Penalized Likelihood Method Exemplified by Scattering and Meteorite Data. *Journal of the American Statistical Association* **75**; 42–73.

Gose, E.E. (1965). An Adaptive Network for Producing Real Functiions of Binary Inputs. *Information Contr.* **8**; 111–123.

Gregor, J. (1969). An Algorithm for the Decomposition of a Distribution into Gaussian Components. *Biometrics* **25**; 79–93.

Gumbel, E.J. (1958). *The Statistics of Extremes*. New York, Columbia University Press.

Habbema, J.D.F., Hermans, J. and van der Broek, K. (1974). A Stepwise Discrimination Program Using Density Estimation. In *Compstat 1974*. Ed. Bruckman, G. Vienna: Physica Verlag, 100–110.

Hajek, J. and Sidak, Z. (1967). *Theory of Rank Tests*. New York: Academic Press.

Hald, A., (1952). *Statistical Theory with Engineering Applications*. New York: Wiley.

Hall, P. (1980). Estimating a Density on the Positive Half Line by the Method of Orthogonal Series. *Annals of the Institute for Statistics and Mathematics* **32**; 351–362.

Hall, P. (1982). Comparison of Two Orthogonal Series Methods of Estimating a Density and its Derivatives on an Interval. *Journal of Multivariate Analysis* **12**; 432–449.

Hall, P. (1983). Orthogonal Series Methods for both Qualitative and Quantitative Data. *Annals of Statistics* **11**; 1004–1007.

Harding, J.P. (1949). The Use of Probability Paper for Graphical Analysis of Polymodal Frequency Distributions. *Journal of the Marine Biological Assoc.* **28**; 141–53.

Hart, J.D. (1985). On the Choice of a Truncation Point in Fourier Series Density Estimation. *Journal of Statistical Computation and Simulation* **21**; 95–116.

Hasselblad, V. (1966). Estimation of Parameters for a Mixture of Normal Distributions. *Technometrics* **8**; 431–444.

Hawkins, J.K., (1961). Self-organizing systems—a Review and Commentary. *Proceedings of the IRE* **49**; 31–48.

Henna, J (1985). On Estimating of the Number of Constituents of a Finite Mixture of Continuous Distributions. *Annals of the Institute of Statistical Mathematics* **37**; 235–240.

Heyde, C.C. (1963). On a Property of the Lognormal Distribution. *J. Roy. Statist. Soc, Ser.. B* **25**; 392–393.

Hildebrandt, E.H. (1931). Systems of Polynomials Connected with the Charlier Expansions and the Pearson Differential and Difference Equations. *Annals of Math. Stat.* **2**; 379–439.

Hills, M. (1969). On Looking at Large Correlation Matrices. *Biometrika* **56**; 249–253.

Hopkins, A. A. and Bond, R. A. (1905). *Scientific American Reference Book.* New York, Munn & Co,

Izenman, I.J. (1991). Recent Developments in Nonparametric Density Estimation. *Journal of the American Statistical Association* **86**; 205–224.

Izenman, I. J. and Sommer, C. J. (1988). Mixtures and Multimodal Densities. Journal of the American Statistical Association **83**; 941–953.

Johannsen, W. (1903). Prof. Johannsen on Heredity (Book Review). *Nature.* December 17, 1903, page 149.

Johnson, N.L. and Kotz, S. (1970). *Continuous Univariate Distributions, Vol. 1.* Boston: Houghton Mifflin.

Jones, M.C. (1991). On Correcting for Variance Inflation in Kernel Density Estimation. *Computational Statistics and Data Analysis* **11**; 3–15.

Kaplan, E.L. and Meier, P. (1958). Nonparametric Estimation from Incomplete Observations. *Journal of the American Statistical Association* **53**; 457–481.

Kendall, M.G. (1943). *The Advanced Theory of Statistics: Vol. 1.* New York: Hafner.

Kendall, M.G. (1952). *The Advanced Theory of Statistics: Vol. 1 Fifth Edition.* New York: Hafner Publishing Company.

Kendall, M.G. (1962). *Rank Correlation Methods.* New York: Hafner Publishing Company.

Kendall, M. and Buckland, W.R. (1960). *A Dictionary of Statistical Terms.* New York: Hafner.

Kizer, K.W. (1989). *Childhood Lead Poisoning in California. Interim Report to the California Legislature.* Chapter 481, Statutes of 1986, Government Code, Section 309.7.

Kline, M. (1953). *Mathematics in Western Culture.* New York: Oxford University Press.

Kline, M. (1972). *Mathematical Thought from Ancient to Modern Times.* New York: Oxford University Press.

Klonias, V. K. (1984). On a class of nonparametric density and regression estimators. *Annals of Statistics* **12**; 1263–1284.

Kronmal, R.A. (1964). *The Estimation of Probability Densities.* Unpublished doctoral dissertation, Division of Biostatistics, University of California, Los Angeles.

Kronmal, R.A. and Tarter, M.E. (1968). The Estimation of Probability Densities and Cumulatives by Fourier Series Methods. *Journal of the American Statistical Association* **63**; 925–952.

Labhart, H. (1947). Ein Auswetegerat fur Elektrophoresdiagramme. *Experientia* **3**; 36–37.

Lackey, R.B. (1972). The Wonderful World of Walsh Functions. In R.W. Zeek and A.E. Showalter, eds. *Proceedings of the Third Walsh Functions Symposium*. U.S. Dept. of Commerse. Springfield, Va. 22151. 2,7.

Lanczos, C. (1956). *Applied Analysis*. Englewood Cliffs, N.J.: Prentice Hall.

Langholz, B. (1984). *Fourier Series Methods in Probability Density Estimation and Distributional Hypothesis Testing*. Unpublished doctoral dissertation, Biomathematics Program, University of Washington, Seattle.

Langholz, B. (1987). Practical Density Estimation Using Fourier Series Methods. Technical Report 11. University of Southern California Dept. of Preventive Medicine.

Langholz, B. (1991). Tests of Distributional Hypotheses With Nuisance Parameters Using Fourier Series Methods. *Journal of the American Statistical Association* **86**; 1077–1084.

Lehmann, E.L. (1959). *Testing Statistical Hypotheses*. New York: Wiley.

Lehmann, E.L. (1963). Asymptotically Nonparametric Inference: An Alternative Approach to Linear Models. *The Annals of Mathematical Statistics* **34**; 1494–1506.

Lehmann, E.L. and D'Abrera, H.J.M. (1975). *Nonparametric Methods Based on Ranks*. Oakland: Holden-Day.

Lindley, D.V. (1965). *Introduction to Probability and Statistics from a Bayesian Viewpoint, Part 2: Inference*. London: Cambridge University Press.

Lipka, J. (1918). *Graphical and Mechanical Computation*. New York: Wiley.

Lock, M.D. (1990). *Optimizing Density Estimates Based on Unweighted and Weighted Mean Integrated Square Error*. Unpublished doctoral thesis, University of California, Berkeley.

Lorentz, G.G. (1953). *Bernstein Polynomials*. Toronto: University of Toronto Press.

Macdonald, P.D.M. (1975). Estimation of Finite Distribution Mixtures. In *Applied Statistics*. Ed. R.P. Gupta. 232–245.

Mardia, K.V., Kent, J.T. and Bibby, J.M. (1979). *Multivariate Analysis*. London: Academic Press.

Medgyessy, P. (1961). *Decomposition of Superpositions of Distribution Functions*. Budapest: Publishing House of the Hungarian Academy of Sciences.

Mickey, M.R. Mundle, P.B. Walker, D.N. and Glinski, A.M. (1963). *Test Criteria for Pearson Type III Distributions*. ARL Technical Documentary Report 63–100. Arlington, Va., Armed Forces Technical Information Agency.

Moore, C.L. (1938). *Summable Series and Convergence Factors*. American Mathematical Society Colloquium Publications, No. 22. New York: American Mathematical Society.

Moore, D.H. (1973). Evaluation of Five Discrimination Procedures for Binary Variables. *Journal of the American Statistical Association* **68**; 339–404.

Muller, H.G. (1984). Smooth Optimum Kernel Estimators of Densities, Regression Curves and Modes. *The Annals of Statistics* **12**; 766–774.

Nadaraya, E. A. (1965). On non-parametric estimates of density functions and regression curves. *Theory Prob. Appl.* **10**; 186–190.

Neyman J. (1959). Contribution to the Theory of the Test. In *Proceedings of the Berkeley Symposium*. Ed. Neyman, J. Berkeley: University of California Press.

Neyman, J. (1959). "Smooth Test" for Goodness-of-fit. *Skandinaviske Aktuarietiddkrift* **20**; 150–199.

Noble, F.W., Hayes, J.E., and Eden, M. (1959). Repetitive Analog Computer for the Analysis of Sums of Distribution Functions. *Proceedings of the IRE* **47**; 1952–56.

Noether, G.E. (1984). Nonparametrics: The Early Years—Impressions and Recollections. *The American Statistician* **38**; 173–178.

Ott, J. (1972). Classification by Orthogonal Functions in Loglinear Models. Unpublished master's thesis, Department of Biostatistics, University of Washington.

Ott, J. and Kronmal, R.A. (1976). Some Classification Procedures for Multivariate Binary Data Using Orthogonal Functions. *Journal of the American Statistical Association* **71**; 391–399.

Padgett, W. J and McNichols, D. T. (1984). Nonparametric Density Estimators from Censored Data. *Communications in Statistics—Theory and Methods* **13**; 1581–1611.

Park, B.U. and Marron, J.S. (1990). Comparison of Data-driven Bandwidth Selectors. *Journal of the American Statistical Association* **85**; 66–72.

Parzen, E. (1962). On Estimation of a Probability Density Function and Mode. *Annals of Mathematical Statistics* **33**; 1065–1076.

Parzen, E. (1977). Nonparametric Statistical Data Science. A Unified Approach Based on Density Estimation and Testing for White Noise. *Statistical Division Report No. 47*, State University of New York at Buffalo.

Parzen, E. (1979). Nonparametric Statistical Data Modeling. *Journal of the American Statistical Association* **74**; 105–131.

Paulson, A.S., and Thornton, J.C. (1975). A New Approach to Goodness-of-fit. *Proceedings of the Sixth Annual Pittsburgh Conference on Modeling and Simulation*, Pittsburgh, Pennsylvania.

Pearson, E.S. and Hartley, H. O. (1956). *Biometrika Tables for Statisticians, Volume 1*. Cambridge: Cambridge University Press.

Pearson, K. (1894). Contributions to the Mathematical Theory of Evolution. *Philosophical Transactions of the Royal Society*, 185A:71–110.

Pearson, K. (1906). Walter Frank Raphael Weldon. 1860–1906. *Biometrika* **5**; 1–52.

Peebles, G. (1940). Some generalizations of the theory of orthogonal polynomials. *Duke Mathematical Journal* **6**; 89–100.

Peterson, A.V. (1977). Expressing the Kaplan-Meier Estimator as a Function of Empirical Subsurvival Functions. *Journal of the American Statistical Association* **72**; 854–858.

Plackett, R.L. (1958). The Analysis of Life Test Data. *Technometrics* **1**; 9–19.

Prakasa Rao, B.L.S. (1983). *Nonparametric Functional Estimation*. New York: Academic Press.

Preston. E.J. (1953). A Graphical Method for the Analysis of Statistical Distributions into Two Normal Components. *Biometrika* **40**; 460–64.

Quandt, R.E. (1958). The Estimation of the Parameters of a Linear Regression System Obeying Two Separate Regimes. *Journal of the American Statistical Association* **51**; 873–880.

Quandt, R.E. (1972). New Approach to Estimating Switching Regressions. *Journal of the American Statistical Association* **67**; 306–310.

Quandt, R.E. and Ramsey, J.B. (1978). Estimating Mixtures of Normal Distributions and Switching Regressions. *Journal of the American Statistical Association* **73**; 730–751.

Ramirez, R.W. (1985). *The FFT-Fundamentals and Concepts*. Englewood Cliffs: Prentice-Hall.

Ray, R. Tarter, M.E., and Lock, M.D., (1991) Mixing Parameter Regression Applied to Groundwater Contaminant Flow. *Computing Science and Statistics-Proceedings of the 21th Symposium on the Interface* (Editor, J. Kettering). American Statistical Association, Alexandria, Virginia.

Rietz, H.L. and Crathorne, A.R. (1924). Simple Correlation. In *Handbook of Mathematical Statistics*, Ed. H.L. Rietz. Boston: Houghton Mifflin.

Ritt, J.F. (1948). *Integration in Finite Terms*. New York: Columbia University Press.

Rosenblatt, M. (1956). Remarks on Some Non-parametric Estimates of a Density Function. *The Annals of Mathematical Statistics* **27**; 832–837.

Rosenblatt, M. (1962). *Random Processes*. New York: Oxford University Press.

Rosenlicht, M. (1975). Differential Extension Fields of Exponential Type. *Pacific Journal of Mathematics* **57**; 289–300.

Rosi, R. and Brunk, H.D. (1988). L_1 and L_2 cross-validation for density estimation with special reference to orthogonal expansions. *Computational Statistics and Data Analysis*, 6:202–227.

Rudemo, M. (1982). Empirical Choice of Histograms and Kernel Density Estimators. *Scandinavian Journal of Statistics* **9**; 65–78.

Rudin, W. (1953). *Principles of Mathematical Analysis*. New York: McGraw-Hill.

Samuelson, P.A. (1943). Fitting General Gram-Charlier Series. *Annals Math. Stat.* **16**; 179–187.

Sarndal, C-E. (1971). Studies in the History of Probability and Statistics.

XXVII - The Hypothesis of Elementary Errors and the Scandinavian School in Statistical Theory. *Biometrika* **58**; 375–391.

Schwartz, S.C. (1967). Estimation of Probability Density by an Orthogonal Series. *Annals of Math Stat* **38**; 1261–1265.

Schwartz, S.C. (1969a). Estimation of Density Functions by Orthogonal Series and an Application to Hypothesis Testing. *Institute of Mathematical Statistics Annual Meeting*, August 19–22, 1969. New York.

Schwartz, S.C. (1969b). A Series Technique for the Optimum Detection of Stochastic Signals in Noise. *IEEE Transactions on Information Theory* **IT–15**; 362–369.

Schweder, T. (1975). Window Estimation of the Asymptotic Variance of Rank Estimators of Location. *Scandinavian Journal of Statistics* **2**; 113–126.

Scott, D. W. (1979). On Optimal and Data-Based Histograms. *Biometrika* **66**; 605–610.

Scott, D. W. (1985). Averaged Shifted Histograms: Effective Nonparametric Density Estimators in Several Dimensions. *Annals of Statistics* **13**; 1024–1040.

Scott, D. W. (1988). A Note on Choice of Bivariate Histogram Bin Shape. *Journal of Official Statistics* **4**; 47–51.

Scott, D. W. (1992). *Multivariate Density Estimation: Theory, Practice, and Visualization*. New York, Wiley.

Scott, D. W. and Factor, L.E. (1981). Monte Carlo Study of Three Data-based Nonparametric Density Estimators. *Journal of the American Statistical Association* **76**; 9–15.

Scott, D. W. and Sheather, S. J. (1985), Kernel Density Estimation with Binned Data. *Communications in Statistics* **14**; 1353–1359.

Scott, D. W., Tapia, R.A. and Thompson, J.R. (1977). Kernel Density Estimation Revisited. *Nonlinear Analysis* **1**; 339–372.

Scheuer, E. M. and Stoller, D. S. (1962). On the Generation of Normal Vectors. *Technometrics* **4**; 278–281.

Shryrock, H.S., Siegel, J.S., Bayo, F., Davidson, M., Demeny, P., Glick, P.C., Grabill, W.H., Grove, R.D., Israel, R.A., Jaffe, A.J., Kindermann, C.R., Larmon, E.A., Nam, C.B. (1973). *The Methods and Materials of Demography*. U.S. Department of Commerce, Washington D.C.

Silverman, B.W. (1978a). Choosing the Window Width when Estimating a Density. *Biometrika* **65**; 1–11.

Silverman, B.W. (1978b). Density ratios, empirical likelihood and cot death. *J. Roy. Statist. Soc. Ser. C* **27**; 26–33.

Silverman, B.W. (1981). Using Kernel Density Estimates to Investigate Multimodality. *Journal of the Royal Statistical Society,* B **43**; 97–99.

Silverman, B. W. (1982). On the estimation of a probability density function by the maximum penalized likelihood method. *The Annals of Statistics* **10**; 795–810.

Silverman, B.W. (1986). *Density Estimation for Statistics and Data Analysis*. London and New York: Chapman and Hall.

Singpurwalla, N.D. and Wong, M.-Y. (1983). Estimation of the Failure Rate—A Survey of Nonparametric Methods. I. Non-Bayesian Methods. *Communications in Statistics—Theory and Methods* **12**; 559–588.

Stanat, D.F. (1966). *Nonsupervised Pattern Recognition Through the Decomposition of Probability Functions*. Technical Report, University of Michigan Sensory Intelligence Laboratory.

Stanat, D.F. (1968). Unsupervised Learning of Mixtures of Probability Functions, In *Pattern Recognition*, Ed. Kanal, L.N. Washington, D.C.: Thompson Book Co., 357–89.

Stone, M. (1974). Cross-Validatory Choice and Assessment of Statistical Predictions. *Journal of the Royal Statistical Society B* **36**; 111–147.

Subrahamian, K.K. (1963). *An Analysis of the Age-distribution at Death of Mice Exposed to Gamma Radiation*. Unpublished M.S. Thesis, Johns Hopkins University.

Szego, G. (1939). *Orthogonal Polynomials*. New York: American Mathematical Society Colloquium Publications, Volume XXIII.

Szigeti, Gy. (1947). Lumeniszkalo Anayagod. *Elegroktechnika* **39**; 70–73.

Sz.-Nagy, B. (1965). *Introduction to Real Functions and Orthogonal Expansions*. New York: University Texts in Mathematical Sciences.

Tapia, R.A. and Thompson, J.R. (1978) *Nonparametric Probability Density Estimation*. Baltimore: Johns Hopkins University Press.

Tarter, M.E. (1975a). A Method for Maximum Likelihood Parameter Estimation Based upon a Reduced Data File. *Proceedings of the Conference on Computer Graphics, Pattern Recognition and Data Structure*. IEEE #75CH0981–1C. 249–252

Tarter, M.E. (1975b). Estimation Procedures for Simulation Applications. In *Perspectives in Biometrics*. Ed. Elashoff, R. Academic Press, Vol. 1, 149–198.

Tarter, M.E. (1975c). Model-free Data Exploration and Pattern Recognition—Discussion and Examples. *Computer Science and Statistics: Proceedings of the 8th Symposium on the Interface*. Ed. Frane, J. Los Angeles: Western Periodicals, 372–381.

Tarter, M.E. (1976). A Graphical Method for Spurious Association Reduction. *Proceedings of the Northwest 76 ACM/CIPS Pacific Regional Symposium*. 87–93.

Tarter, M.E. (1978a). Implementation of Harmonic Data Analysis Procedures. *Computer Science and Statistics: Proceedings of the 11th Symposium on the Interface*. Ed.Gallant, A. R. and Gerig, T. M.. North Carolina State University: Institute of Statistics. 234–239.

Tarter, M.E. (1978b). Interactive Graphical Isolation of Homogeneous Data Subgroups. *Computer Programs in Biomedicine* **8**; 81–86.

Tarter, M.E. (1978c). A Description of the GRAFSTAT Clinical Life Table

Options. *American Statistical Association 1978 Proceedings of the Statistical Computing Section*, 44–54.

Tarter, M.E. (1979a). Trigonometric Maximum Likelihood Estimation and Application to the Analysis of Incomplete Information. *Journal of the American Statistical Association* **74**; 132–139.

Tarter, M.E. (1979b). Biocomputational Methodology—An Adjunct to Theory and Applications. *Biometrics* **35**; 9–24.

Tarter, M.E. (1980). Comment on Density Estimation and Bump-Hunting by Penalized Likelihood Method Exemplified by Scattering and Meteorite Data by Good and Gaskins. *Journal of the American Statistical Association* **75**; 63–65.

Tarter, M.E. (1981). Modular Implementation of Graphical Curve Estimation Procedures. *American Statistical Association 1978 Proceedings of the Statistical Computing Section*, 69–78.

Tarter, M.E. (1983). On the Use of Linear Harmonic Forms to Simultaneously Generalize Certain Statistical Computations. *Computer Science and Statistics: The Interface* (Ed. K.W. Heiner, R.S. Sacher, and J.W. Wilkinson). New York: Springer-Verlag, 232–241.

Tarter, M. E. (1986). Mean integrated squared error sampling. *Journal of the American Statistical Association* **81**; 234–242.

Tarter, M. E. and Clark, V. (1965). Properties of the Median and Other Order Statistics of Logistic Variates, *The Annals of Mathematical Statistics* **36**; 1779–1786.

Tarter, M. E., Cooper, R. C. and Freeman, W. R. (1983). A Graphical Analysis of the Interrelationships among Waterborne Asbestos, Digestive System Cancer and Population Density. *Environmental Health Perspectives* **53**; 79–89.

Tarter, M.E. and Freeman, W.R. (1985). On Curve Estimation and Pattern Recognition in the Presence of Incomplete Data. *Computer Science and Statistics: The Interface*, (Ed. Gentle, J.), New York: North-Holland Publishing Company, 191–198.

Tarter, M.E. and Freeman, W. (1986). On Graphing Estimated Distributions Using Modified Scatter Diagrams. *Computational Statistics and Data Analysis* **7**; 51–67.

Tarter, M.E. and Freeman, W. (1987). A Graphical Procedure for Distinguishing between Two Data Analysis Pitfalls. *Communications in Statistics: Simulation and Computation* **16**; 427–437.

Tarter, M.E., Freeman, W.R. and Hopkins, A. (1986). A FORTRAN Implementation of Univariate Fourier Series Density Estimation. *Communications in Statistics, Part B:Simulation and Computation* **15**, (3), 855–870.

Tarter, M.E., Freeman, W. and Polissar, L. (1990). Applications of Modular Nonparametric Subsurvival Estimation. *Journal of the American Statistical Association* **85**; 29–37.

Tarter, M.E., Holcomb, R. and Kronmal, R.A. (1967). A Description of New

Computer Methods for Estimating the Population Density. *Proceedings of the ACM* **22**; 511–519.

Tarter, M.E. and Kowalski, C. (1972). A New Test for and Class of Transformations to Normality. *Technometrics* **14**; 735–743.

Tarter, M.E. and Kronmal, R.A. (1970). On Multivariate Density Estimates Based on Orthogonal Expansions. *The Annals of Mathematical Statistics* **41**; 718–722.

Tarter, M.E. and Kronmal, R.A. (1976). An Introduction to the Implementation and Theory of Nonparametric Density Estimation. *American Statistician* **30**; 105–112.

Tarter, M.E. and Lock, M.D. (1991). Model-free Curve Estimation: Mutuality and Disparity of Approaches. *Journal of Official Statistics* **17**; 58–73.

Tarter, M.E. Lum, S. B. and Dinaburg, A. (1977). A Brief Example of Interactive Nonparametric Regression Applied to Ballistocardiogram Data. *Proceedings of Medinfo 77*, Second World Conference on Informatics.

Tarter, M.E. and Marshall, J. (1978). A New Procedure Designed for Simulation Studies. *Communications in Statistics- Series B* **7**; 283–93.

Tarter, M.E. and Raman, S. (1971). A Systematic Approach to Graphical Methods in Biometry. *Proceedings of the Sixth Berkeley Symposium on Mathematical Statistics and Probability. Volume IV: Biology and Health,* Ed. Neyman, J. Berkeley: University of California Press, 199–222.

Tarter, M.E., Rigsbee, E.O. and Wong, J.T. (1976). Interactive Editing of Biomedical Data. *Computer Programs in Biomedicine* **6**; 117–123.

Tarter, M.E. and Silvers, A. (1975). Implementation and Applications of Bivariate Gaussian Mixture Decomposition. *Journal of the American Statistical Association* **70**; 47–45.

Taylor, A.E. (1955). *Advanced Calculus*. Boston: Ginn and Co.

Thomas, D.R. and Pierce, D.A. (1979). Neyman's Smooth Goodness-of-fit Test when the Hypothesis is Composite. *Journal of the American Statistical Association* **74**; 441–445.

Tiku, M.L. (1968). Estimating the Parameters of Normal and Logistic Distributions from Censored Samples. *Australian Journal of Statistics* **10**; 64–74.

Titterington, D. M., Smith, A. F. M. and Makov, U. E. (1985). *Statistical Analysis of Finite Mixture Distributions*. New York: Wiley.

Tolstov, G.P. (1962). *Fourier Series*. Englewood Cliffs New Jersey: Prentice-Hall.

Tukey, J.W. (1965). The Technical Tools of Statistics. *The American Statistician* **19**; 23–36.

Tukey, J.W. (1974). Discussion of the Kronmal Tarter Chapter. *Exploring Data Analysis, The Computer Revolution in Statistics*. (Eds. Dixon, W.J. and Nicholson, W.L.) Berkeley: University of California Press, 397–400.

Tukey, J.W. (1977). *Exploratory Data Analysis*. Reading Mass.:Addison-Wesley.

Uesaka, Y., (1971). Construction of a Complex-Valued Nonlinear Discriminant Function for Pattern Recognition. *Institute of Electrical and Electronic Engineers Transactions on Systems, Man and Cybernetics* **1**; 194–215.

Wahba, G. (1977). Optimal Smoothing of Density Estimates. In *Classification and Clustering,* Ed. Van Ryzin, J. New York: Academic Press, 423–458.

Wahba, G. (1981). Data-based Optimal Smoothing of Orthogonal Series Density Estimates. *The Annals of Statistics* **9**; 146–156.

Wald, A. (1947). *Sequential Analysis*. New York: Wiley.

Walsh, J.L. (1923). A Closed Set of Normal Orthogonal Functions. *American Journal of Mathematics* **1**; 5–24.

Wand, M.P., Marron, J.S. and Ruppert, D. (1991). Transformations in Density Estimation. *Journal of the American Statistical Association* **86**; 343–352.

Watson, G.S. (1964). Smooth regression analysis. *Sankhya: Series A* **26**; 359–372.

Watson, G.S. (1967). Another Test for the Uniformity of a Circular Distribution. *Biometrika* **54**; 675–677.

Watson, G.S. and Leadbetter, M.R. (1963). On the Estimation of the Probability Density, I. *Annals of Math. Stat.* **34**; 480–491.

Watson, G.S. (1969). Density Estimation by Orthogonal Series. *The Annals of Mathematical Statistics* **40**; 1496–1498.

Wegman, E. J. (1970) Maximum likelihood estimation of a unimodal density function. *The Annals of Mathematical Statistics* **41**; 457–471.

Wegman, E. J. (1970) Maximum likelihood estimation of a unimodal density, II. *The Annals of Mathematical Statistics* **41**; 2169–2174.

Weichselberger, V.K. (1961). *Ü* ber ein Graphisches Verfahren zur Trunnung von Mischvertielungen und zur Identifikation Kupietier Normal-Verteilungen bei Grossem Stichprobennumfang. *Metrica* **3**; 178–229.

Weldon, W.F.R. (1893). On Certain Correlated Variations in *Carcinus moenas*. *Proceedings of the Royal Society of London* **54**; 318–329.

Weldon, W.F.R. (1901). The Scope of *Biometrika*. *Biometrika* **1**; 1–2.

Whittle, P. (1958). On the Smoothing of Probability Density Functions. *Journal of the Royal Statistical Society, B* **20**; 334–343.

Wilk, M.B. and Gnanadesikan, R. (1964). Graphical Methods for Internal Comparisons in Multiresponse Experiments. *The Annals of Mathematical Statistics* **35**; 613–631.

Wilks, S.S. (1962). *Mathematical Statistics*. New York: John Wiley and Sons.

Winter, B.B. (1979). Convergence rate of perturbed empirical distribution functions. *J. Appl. Prob.* **16**; 163–173.

Wolfowitz, J. (1942). Additive Partition Functions and a Class of Statistical Hypotheses. *Annals of Math. Stat.* **13**; 247–279.

Yakowitz, S. J. and Szidarovszky, F. (1985). A comparison of kriging with nonparametric regression methods. *Journal of Multivariate Analysis* **16**; 21–53.

Zippin, C. and Armitage, P. (1966). Use of Concomitant Variables and Incomplete Survival Information in the Estimation of an Exponential Survival Parameter. *Biometrics* **22**; 665–672.

Author Index

Subject Index